THE NEW NATURALIST

A SURVEY OF BRITISH NATURAL HISTORY

THE WORLD OF THE SOIL

The aim of this series is to interest the general reader in the wild life of Britain by recapturing the inquiring spirit of the old naturalists. The Editors believe that the natural pride of the British public in the native fauna and flora, to which must be added concern for their conservation, is best fostered by maintaining a high standard of accuracy combined with clarity of exposition in presenting the results of modern scientific research. The plants and animals are described in relation to their homes and habitats and are portrayed in the full beauty of their natural colours, by the latest methods of colour photography and reproduction.

THE NEW NATURALIST

THE WORLD OF
THE SOIL

by

SIR E. JOHN RUSSELL

D.Sc., F.R.S.

LATE DIRECTOR OF
THE ROTHAMSTED EXPERIMENTAL STATION

WITH 4 COLOUR PHOTOGRAPHS
44 PHOTOGRAPHS IN BLACK AND WHITE
AND 11 TEXT FIGURES

COLLINS
ST JAMES'S PLACE, LONDON

First published, 1957
Second Edition, 1958
Third Edition, 1963
Fourth Edition, 1967
Fifth Edition, 1971

ISBN 0 00 213255 9

© *Sir E. John Russell,* 1957

Printed in Great Britain
Collins Clear-Type Press: London and Glasgow

CONTENTS

PLATES IN COLOUR

1a   Iron deficiency: sweet cherry leaves   18
1b   Calcium deficiency: tomatoes   18
2a   Magnesium deficiency: apple leaves   146
2b   Manganese deficiency: pea seeds   146

PLATES IN BLACK AND WHITE

I   Old and new laboratories at Rothamsted   50
II   First stage in soil formation   51
IIIa Clay particles x 40,000   66
IIIb Coating on soil particles, Rothamsted   66
IVa, b Bacteroids of Rhizobium from Red Clover   67
IVc   Lucerne plants with and without their proper bacteria   67
Va   Conidial head of *Aspergillus niger*   98
Vb   Bacteriophage, x 56,500   98
VI   Minute soil animals   99
VII   Meadow nematode, x 900   114
VIII Section of tomato roots infested by nematodes   115
IXa, b Boron essential for nodule development in leguminous
       plants   162
IXc   Molybdenum deficiency   162

It should be noted that throughout this book Plate numbers in Arabic figures refer to Colour Plates, while Roman numerals are used for Black-and-White Plates

vii

Plates

vi

LINE DIAGRAMS

EDITORS' PREFACE

SUCCESSIVE VOLUMES in the *New Naturalist* series have dealt with life in the open ocean, on the ocean margins, on land, in the air, and in fresh water. There remains an environment which is peculiarly distinct and which is tenanted by a remarkable range of animals and plants of its own. It is indeed a world on its own — the world of the soil. Many, indeed the great majority, of the inhabitants of this world never leave it: they play an important part in the life of those higher organisms which use the soil for rooting, the collecting of sustenance or perhaps simply for the siting of convenient homes. Too often in the past the soil has been regarded as simply a superficial layer over the land, annoying to the geologist because it hides the rocks he seeks to study; important to the botanist as the medium in which the plant root systems develop; vital to the agriculturist as affecting the yield of crops he has selected as of economic importance. In recent years has grown up the study of pedology, or soil science, which seeks to study and classify soils, to describe them, name them, and trace their development, but the pedologist is only concerned incidentally with the creatures which make the soil their permanent home.

This book presents a different picture. It gets inside this strange world and analyses its structure and how it has developed; the circulation of water and gases which give the soil an atmosphere and a succession of climates quite different from those of the air above. It deals with the varied forms of life made possible or impossible by the varied conditions and so to a concept of the soil providing habitats of differing and frequently contrasted characters. It may be truthfully said that Sir John Russell has devoted his long scientific life to the soil. In his recently published autobiography, aptly entitled *The Land Called Me,* he shows how he, a chemist by training, saw his life's work in the application of science to developing the productivity of the soil in the service of man. His classic *Soil Conditions and Plant Growth* has gone through many editions, evolving in line with the evolution of soil science; under his 30 years as Director of the Rothamsted Experimental

Station, that famous institution stepped from just one of many agri-cultural experimental stations to one which pioneered in innumerable directions, not least in soil science, so as to become the centre of the Commonwealth Soil Bureau.

Sir John and his former colleague and predecessor at Rothamsted, Sir Daniel Hall, laid a firm foundation for subsequent studies as early as 1911, with the publication of their *Agriculture and Soils of Kent, Surrey and Sussex*. In the intervening years no one has followed more closely the development of soil studies yet never losing sight, as so many other have done, of the wider implications of detailed work. No one certainly could have written such a fascinating account of this brave new world as the Editors now take pride and pleasure in presenting to the readers of the *New Naturalist* series.

THE EDITORS

AUTHOR'S PREFACE

AUTHOR'S PREFACE

THIS BOOK has been written for those thoughtful people who, though not scientific specialists, are nevertheless interested in the countryside and in their gardens, and who wish to read a little more in "Nature's infinite book of secrecy". The soil is among Nature's greatest marvels. A clod of earth, seemingly simple and lifeless, is now known to be highly complex in structure, its particles most elaborate in their composition, with numerous invisible crevices inhabited by prodigious numbers of living organisms inconceivably small, leading lives of which we can form only the haziest conception, yet somehow linked up with our lives in that they produce the food of plants which constitute our food, and remove from the soil substances that would be harmful to us. It would be flattering to our self esteem to think that they are there for that purpose, but it is safer to take the wider view and to suppose that they, like ourselves, are leading their own lives, the purpose of which is completely hidden from us.

Fifty years ago this book would have been much easier to write than it has been today. Enough was then known about the wonders of the soil to show that deeper mysteries lay beyond, the facts gleaned were simple, the generalisations were broad and easily comprehended, and there was abundant scope for imaginative and picturesque presentation.

Now it is all very different. Experts in many scientific subjects have spent years in studying the soil, and vast numbers of learned memoirs have been written about it. Geologists have discussed at length its mode of formation, chemists have investigated the wonderful structure of its minerals and have spent years in trying to unravel the complexities of its organic constituents, while biologists of many different kinds have laboured at great length to find out what they could about the myriad forms of life invisible to our eyes that live in its mysterious recesses. The work still goes on; at the Rothamsted Experiment Station alone something like a hundred highly competent scientists aided by fully trained technicians and armed with all kinds of complex appliances are strenuously engaged in studying the soil from their different points of view; there are many other investigators in this country and still more in the lands overseas. Yet each worker would admit that he or she is only on the fringe of the subject: what is still unknown is immeasurably greater than what has already been discovered. We are forced to the conclusion that we can never hope to know all about this apparently simple lifeless clod of earth.

I have tried to set out the broad outlines of the picture as we see it at present, avoiding as far as possible the dangerous pitfalls of oversimplification and of imaginative but inaccurate detail. Throughout I have kept in mind the educated but non-specialist reader and I have used only a minimum of technical terms: at the same time I have sought to make the book useful to the student and to avoid offending the expert by inaccuracies resulting from efforts to simplify. Nearly all my working life has been devoted to the subject, but it has grown to such vast dimensions that no one person can hope to comprehend all of even our present very limited knowledge. Each section of the book has been read and criticised by the appropriate expert: this help is acknowledged in the text but I must express my special thanks to my friends at Rothamsted who have ungrudgingly come to my aid whenever I found myself in difficulties.

I am indebted to the Controller of H.M. Stationery Office for permission to reproduce the four colour illustrations, which appeared in *The Diagnosis of Mineral Deficiences in Plants* by Professor T. Wallace.

E. JOHN RUSSELL

February, 1957.   *Woodstock, Oxon*

UNITS OF MEASUREMENT

Many of the objects described in the following pages are so small that ordinary English units are quite unsuitable. The following are used instead:

10,000	Ångstrom units (Å)	=	1 micron (μ)
1,000	millimicrons (mμ)	=	1 micron
1,000	microns	=	1 millimetre (m.m.)
25·4	millimetres	=	1 inch
1,000	millimetres	=	1 metre
1	metre	=	39·37 inches
1,000	cubic centimetres (c.c.)	=	1 litre
1	litre	=	1·76 pints

CHAPTER I

HOW SOIL WAS FORMED: THE
MINERAL SKELETON

In one of the earliest English books on the soil, called *Terra*, John Evelyn in 1675 states that the theorists of his day "reckon up no fewer than one hundred seventy nine millions one thousand and sixty different sorts of Earths". I shall not argue this question. Many farmers could be found to assert that no two fields on their farm had similar soils, and that even on the same field there were several different kinds. Many gardeners would say the same.

Nevertheless the different soils have a great deal in common. The soils of the world can be classified in a few great groups: most British soils fit into two or three of these. Every soil, whatever its group may be, is composed of a mineral part which forms its skeleton, and an organic part which gives it some distinctive attributes; each has certain characteristics common to all the members of the group. The differences between the members lie in the relative proportions of the constituents and in the differences in their chemical composition.

In the first place it is necessary to be clear about what is meant by the word "soil". It is used in two or three different senses. The civil engineer means all disintegrated material lying on top of solid rock: it may be many feet in thickness. The pedologist means the layer in which chemical and physical changes have taken place; while the agriculturist and the gardener generally mean the top six or nine inches of earth in which their plant roots grow, a layer which is easily distinguished as a rule from the underlying material because of its difference in colour and structure. The underlying layer is called the subsoil, and the two together, surface and subsoil, roughly add up to the pedologist's more precise term. The surface soil is by far the more important as a habitat for plants and other living things, but the subsoil serves many useful purposes so it is included in this book.

A large part of the difference between the various soils of this country is due to the difference in the rocks from which they are formed. After our globe had solidified forming a greatly corrugated surface of mountain and valley, dry land and seas, the rain, the heat and the frost caused fragments to split off from the rocks and they were washed down into the valleys or into the seas where they mingled with remains of sea life; there in course of ages they formed layers of sediment so thick that their weight consolidated them into new rocks. Later earth movements raised these out of the water and converted them into dry land: once again the round of disintegration and decomposition set in, and the particles were carried down to the new seas, again to mingle with residues of sea life and form thick deposits which ultimately once more became rocks, were again thrown up as land, and again subjected to denudation.

Much of England has been twice submerged, and some of it four times or more, each submergence resulting in the formation of a great stratum of new rock from fragments washed down from the higher unsubmerged rocks or built up of the residues of sea animals. The strata lie one on top of the other: had they remained in the position in which they were first formed England would have been a rather monotonous plain except where some of the older hills stood out. But they did not; there has been considerable dipping towards the east, tilting up the strata and leaving them exposed in such a way that the softer ones were eroded by frost and rain into valleys while the harder more resistant ones remained as higher ground.

So it happens that you can start on the east coast where the North Sea is in some places eroding away the land and in others depositing new material; going westward you pass over the newest formations, you traverse the chalk formation deposited while the land lay under the Cretaceous sea; this rises to the chalk escarpment. Then you drop down to the clay valley of the Gault, then rise to the low escarpment of the Lower Greensand, then drop to the clays deposited in the Jurassic Sea. Further on the underlying Oolite limestones come to the surface and the land rises through breezy uplands to the superb escarpment which, beginning in the Gloucestershire Cotswolds, runs north easterly across England, with many famous view points: Birdlip, Sunrising Hill, Edge Hill, Rockingham, through Rutland and Lincolnshire to East Yorkshire where it forms the moor country about Goatland and ends at the coast between Scarborough and Whitby. Beyond the escarpment come the lower lying undulating Lias clays and limestones through which flow the Severn, the Stratford Avon, and many charming and peaceful streams in some of the most famous vales in England: broad belts of pasture, noted for generations for grazing and historic fox hunts.[1] Further west come the red sandstones and pebble beds of the Trias formations: these were formed in inland seas and in deserts at a time when England was a hot, dry land; beyond again come the Coal Measures, the remains of vegetation long since extinct, still further, and geologically older, lies another stretch of red soils, the Devonian or Old Red Sandstone. This is the end of lowland Britain. The mountain lands include the Millstone grits, Carboniferous Limestone and igneous rocks of the Pennines, and further north in Wales the Silurian, Ordovician and Cambrian; during all the changes recorded above they were never submerged, but in their long exposure to the weather they have become rounded by erosion; the tilting of the rocks however has produced many steep slopes and scarps. (Plate II. p. 51).

Each of these formations has its own types of soil with characteristics which go back to their beginnings. And the times have been long. Something like 500 million years have elapsed since the Cambrian group of rocks began to form, some 150 million years since the Oolite group were being laid down, and probably 120 million years since the formation of the great Chalk deposits—the last to make a dominating feature in the English landscape. Yet in all those incomprehensible periods of time the differences in character resulting from the differences in conditions of formation have never been smoothed out: an Oxford Clay soil still differs from an oolite cornbrash or clay although they have been side by side for millions of years; their main characters are as nearly permanent as anything on this earth can claim to be.

British soils as they stand today are not necessarily related to the rocks over which they lie. Profound changes were effected during the ages when a change of climate caused most of Britain north of the Thames to become covered with ice. There is much uncertainty about the details, but there appear to have been four ice ages separated by warmer periods. The first began some 500,000 years ago and the last petered out in England about 15,000 years ago—in Scotland there

[1] The belt near the escarpment forms the very interesting floral frontier beyond which some of the wild plants of the south, such as the crimson flowered stemless thistle, *Cirsium acaule*, do not pass northward or, with rare exceptions, westward; and some of the northern and western plants, like the little alpine-arctic fern, *Asplenium viride*, do not pass southwards.   (A. R. Clapham).

were glaciers in some valleys 4 or 5 thousand years later. The ice profoundly affected the British landscape. The glaciers as they moved carved out valleys, ground down to a powder some of the rocks over which they passed, transported rocks, stones and fine ground rock material from one district to another. Then about 10,000 B.C. a warmer climate set in, the ice melted and the water rushing away carried with it much of the finer material leaving behind sands, gravel, and stones. The ice cap of the third glaciation ended north of the Thames as that of the second had done: it passed through Essex, Hertfordshire, and counties to the West. The jumble of soils in Hertfordshire—gravel here, clay there, a stretch of sand somewhere else—is the result of the way in which the frozen sludge underlying the glaciers or pushed on by them got sorted out by rain and streams from the ice when it melted. The fourth glaciation ended on the York-Severn line.

When finally the ice had gone England must for a long while have been a very slushy muddy place with a frozen subsoil, the permafrost layer, but gradually it became drier and vegetation spread. The first plants covering the ground while it was still cold and wet were of the tundra type: mosses and small very hardy willows and birches. Some of these plants still survive in the mountains of Wales and the North of England, notably *Betula nana, Salix herbacea, Dryas octopetala, Thalictrum alpinum* and *Oxyria digyna;* also in the Snowdon area *Armeria maritima* and the fern *Cryptogramme.*[1] Later on conditions became warm enough for trees and they came in from the rest of the Continent—for England was not yet an island—and gradually moved northwards; first birch then pine, then hazel and later elm and oak. Still later, about 5000 B.C. conditions became moister, lime and elder came in and the forest growth was more mixed; some of the forest on the higher ground became bog and as the trees died and fell their remains were covered up with peat which has preserved some of them to this day. Elsewhere oak-alder forest was wide-spread with holly, ivy, lime and ash. It was at the opening of this period that the most momentous event in our history occurred: the low lying stretch of land that separated us from the Continent was finally submerged and Great Britain became an island.

Paleolithic hunters had already been here, but about 2,500 B.C. Neolithic man came in bringing cereals and their weeds with him. Towards the end of this period beech began to extend in S.E. Britain and to move northwards.

[1] H. Godwin, *History of the British Flora,* Cambridge Univ. Press, (1956.).

After about 500 B.C. and up to the present time the conditions became wetter and cooler; the tree line moved down hill in the western highlands and the peat bogs increased in size and in number; the peat plants became more common.

The story of the colonisation of our islands by plants is one of the most fascinating in the whole realm of science; it has been a fine piece of detective work made possible by the study of pollen grains found at different depths in peat bogs, and by dating accompanying fragments of plant tissues by measuring the stage of decay of their radioactivity. Much of this work has been done by Dr. H. Godwin and his colleagues at Cambridge.[1]

Throughout these various changes the country as a whole remained covered with forest, dense over large areas, the tree line moving up and down the hills as the climate changed, but there was scrub and heath vegetation on the drier sands and thin chalk where prehistoric man had deforested them, bogs and marsh vegetation in the low lying water-logged areas, and bog vegetation in the high districts above the tree level. Much of the lowland forest in the southern part of England had been cleared in Roman times: if we accept Sir Mortimer Wheeler's estimate of a population of 1½ millions some 2 million or more acres of land must have been in cultivation to provide them with food. Much of this doubtless went back to scrub and forest in the unsettled period following the Roman withdrawal, but later on clearing was resumed. Some of the forests persisted right through to historic times, there are still remnants in places: Epping and Sherwood Forests are much modified examples. Most of our soils have been under forest for some thousands of years and only relatively small areas have been cleared for more than about 1,500 years, which has not been enough to eliminate the profound effects the forest had on them.

The ice had begun the formation of our soils by grinding up the rock fragments and transporting them; as it melted the water and the rain sorted out the particles and left the framework of the soil much as it is today, but the forest had introduced a new element that completed the formation process: the organic matter, which being a source of food and of energy, made possible the establishment of a whole population of living organisms that carry on their vital processes within the soil and in so doing make it a suitable habitat for plants.

[1] It is described in his *History of the British Flora.*

The simplest method of studying the soil is to deal separately with the mineral and the organic constituents. The mineral constituents are sorted out according to the size of the particles: this is done by a modification of a process that can be seen at work in any hill stream. Quickly flowing water carries away the fine material leaving the coarse particles to settle forming a sandy or stony bed according to the speed of the water; as the stream widens and the rate of flow slackens some of the fine material is deposited forming a muddy bed on which water plants can grow and by blocking the flow increase still further the rate of deposition. The process is very marked in some of the lakes made by the great landscape gardeners of the 18th and early 19th Centuries, where constant cleaning was needed to prevent silting up—now it often goes on unchecked. Some of the material, however, is so fine that it does not settle but is carried forward into the river and on to the sea. There a new factor comes into play. The salt in the sea-water causes some of this fine material to set into flocks which settle, carrying with them some of the fine material which does not flocculate, and in time forming the bar which is often such a nuisance to navigation.

The laboratory procedure is similar in principle but elaborated to greater refinement. The soil is first spread out to dry, then the lumps are broken by gentle pounding and the material is passed through a sieve of 2 m.m. mesh: whatever fails to pass is classed as gravel and stones. Soil is thus defined as material the particles of which do not exceed 2 m.m. diameter; this is not an entirely arbitrary figure: there are good and sufficient reasons for it. The particles, however, are not all uniform; some are aggregates of smaller particles and they have to be broken down by gentle chemical treatment; this done there is further sieving through a sieve of 0·2 mm. mesh and the material passing through is separated into fractions either by a stream of water, the method favoured on the Continent, or, as in this country, by letting it fall through a column of water at standard temperature (usually 20°C) and collecting the fractions falling through a specified height in a specified time. This serves for particles above ·002 mm. diameter but not for those smaller; they have to be flocculated by a process similar to the formation of a river bar. The fractions are accurately specified by the rate at which they settle,[1] but a better mental picture is obtained

[1] denoted by pv, the negative logarithm: a particle of settling velocity 10^{-6} cm per sec. has a pv of 6.

by specifying the diameters the particles would have if they were spherical and had a specific gravity of 2·6. Neither assumption is quite correct: even the finest, the clay particles, are irregular in shape (Plate III) but nothing better has yet been devised, and 2·6 is a fairly good average figure. The conventional sizes for the chief fractions and their usual distribution in British soils is as follows:—

EQUIVALENT DIAMETER, MM.	2–0·2	0·2–0·02	0·02–0·002	Below 0·002
NAME OF FRACTION	Coarse sand	Fine sand	Silt	Clay
FREQUENT RANGE: per cent				
Sandy soils	Above 70			Below 20
Loams	70 — 40			20 — 40
Clay soils	Below 40		30 — 40	Above 40

These large divisions are further subdivided by soil surveyors.

Throughout the lowlands of Britain the particles composing the soil have suffered so many changes for so long that only the most resistant have been able to survive as sand or silt; those capable of decomposition have been worn down to much smaller size. The most resistant of all minerals is quartz—silica, SiO_2: it cannot be further decomposed except in a laboratory where it can be resolved into its elements, but the silicon easily and rapidly oxidises to form silica once more. It is one of Nature's end products, and there need be no surprise that while parent rocks contain on the average only about 12 per cent of quartz, the sand and silt fractions in British lowland soils commonly contain 90 to 95 per cent: all the other material has gone, this only remains indestructible. The wide range of particle size between coarse sand and silt may be due to mechanical action such as grinding by the glaciers, but is not likely to be due to solution. The quartz contributes nothing whatever to the nutrition of plants or of any of the other living things in the soil. But its indestructibility makes it an extremely valuable component of the skeleton of the soil. Some upland soils which are derived direct from ancient or igneous rocks and have not been subject to such drastic weathering as those in the lowlands still retain much of the parent rock in the sand and silt fractions and while this may be less resistant than quartz it is equally useless to plants and other soil inhabitants.

The sand particles are irregular in shape, and this gives them one of their useful properties: they do not fit compactly together in the soil: considerable spaces are left between them, each small but so numerous that the aggregate space is considerable. It seems incredible but nevertheless it is true that in an apparently solid clod of earth only about half is usually solid matter, the other half is simply empty except for the air and water it contains.

Silt particles may cause trouble in a soil by blocking up the pore spaces and so impeding drainage. Like the sand particles they are chemically inert and contribute nothing to the food of plants or any other members of the soil population. Between the fine silt and the coarse sand there are no important fundamental differences, but a steady gradation of properties consequent on the diminishing size of the particles. This part of the soil is concerned almost entirely with its architecture and its physical properties, not at all with its chemical reactivity.

A remarkable change sets in, however, on passing to the clay fraction, the particles with equivalent diameter less than 0·002 mm. In the mass it is sticky when wet, becomes hard and horn-like on drying, swells when wetted, shrinks on drying. When stirred up in pure water (distilled, or clean rain water) it clouds the liquid but is rapidly clotted on addition of lime water or a solution of salt or an acid, but it remains persistently in suspension on addition of an alkali; it is chemically active and can take part in some very important changes. To some extent these properties are the result of the enormous increase in the total surface of the particles in consequence of their small size. As is well known, the total surface of the particles of a given quantity of a solid increases rapidly as the particles become smaller. The illustration usually given is that of a cube: so long as it remains whole its six sides represent its total surface, but as it is cut up into smaller parts new surfaces continually appear: the mathematically minded may work out the total surface that would be attained if a cube of one inch side were reduced to the dimensions first of sand particles, then of clay particles.

The total surface of even a small weight of soil is unexpectedly large. J. R. H. Coutts has calculated that one ounce of a Cruden Bay soil, a silty clay loam, has a specific area of nearly 400 square feet: this is a minimum value because it assumes that the particles are spheres; they are not and hence the true value is likely to be higher. But this figure pales into insignificance beside the value of more than a quarter of a million square feet—more than six acres—for an ounce of the clay contained in the Rothamsted soil. The surface soil contains 20 per cent of this, and the lower depths up to 40 per cent. This extensive surface of the clay particles brings into prominence certain glue-like "colloidal" properties imperceptible in the sand fractions and hardly shown by the silt: these include the power of sticking to other substances and also to various chemical elements or groups—ions—and holding them so that they are not easily washed off.

The clay minerals are the most interesting constituents of the clay fraction; they are derived from the original igneous or metamorphic rock by weathering[1] but differ chemically from the parent material in being more reactive. They are complex in composition but they have some of the characteristics of a simple salt in that they can exchange bases with simple salts and the base so taken up cannot be removed by mere leaching but can only be dislodged by another base or a hydrogen ion. Chemists had made many efforts to unravel their constitution, but it was the development of X-ray photography that first revealed their structure. The crystals in the clay minerals are very small: more than 25,000 of them would go to an inch, and to make photographs showing their internal structure has been one of the triumphs of modern science.[2]

As invariably happens their structure turned out to be much more complex but also far more wonderful than had at first been supposed. The minerals in British soils fall mainly into two groups known respectively as the mica and the kaolin types. In each group the units are built up of sheets of silica and of hydrated alumina molecules SiO_2 and $Al(OH)_3$. In the mica type each alumina sheet is set between two silica sheets, so that the unit can be likened to a sandwich: these are piled one on top of the other to form the particle. In the kaolin group the alumina sheet is on top of the silica sheet, like butter on a slice of bread: the particle is made up of these slices piled on each other as in the mica group. The groups can be called 2:1 and 1:1 respectively.

The units are extremely thin: so thin indeed that neither inches nor millimetres are suitable for measuring them, and the Ångstrom units

[1] Some are synthesized in the soil and some derived direct from the parent material but the products of weathering are by far the most important.

[2] Much of the pioneering work in this country was done by W. L. Bragg and is recorded in his book, *The atomic structure of Minerals* (Cornell Univ. Press). Modern applications to soil studies are made by the Pedology Department at Rothamsted, and by the staff of the Macaulay Institute for Soil Research, Aberdeen.

are used instead: 1 Å is equal to 10^{-8} cms: there are 250 millions of them in one inch. For some of the kaolin minerals the distances between a sheet in one layer and the corresponding sheet in the next may be only 7 Å., for some members of the mica group which have an accordion-like quality of expansion and contraction the range may be from 10 Å to 50 Å or more.

Complications arise from several causes. The atoms do not lie on a flat surface, but are grouped in three dimensions: those of silica in a tetrahedron with the silicon at the centre and an oxygen atom at each corner, and those of alumina as an octahedron with the aluminium atom at the centre and six oxygen atoms at the corners. Three of the oxygen atoms of the tetrahedron are shared with other silica tetrahedra so building up a sheet or lattice layer; similarly with the alumina. During the formation of the minerals there has been a certain amount of replacement of one element by another, without, however, greatly disturbing the pattern; this is called isomorphous replacement. Some silicon was replaced by aluminium, some aluminium by iron, magnesium, or occasionally other metals.

All these atoms carry electric charges: those acting as bases, the cations, have positive charges, while the acidic groups, the anions, have negative charges, the numbers of charges being proportional to the valencies. Isomorphous replacement alters the electrical status of the lattice layers because it is usually by an element of lower valency, i.e, fewer charges: quadrivalent silicon is replaced by trivalent aluminium, and trivalent aluminium by divalent magnesium; the result is a decrease in the number of positive charges in the lattice and a development of negative charges on the surface.[1] These charges attract positively charged ions and hold them till they are replaced by others: it is in this way that base exchange comes about.

It is unfortunate that the three dimensional structure of the clay minerals cannot be illustrated by diagrams; the only way of doing it is to set up models out of balls fitted with rods that can join together, the monovalent hydrogen having one, the divalent oxygen two, the trivalent aluminium three, and the quadrivalent silicon four.

In the typical mica one quarter of the silicon atoms are replaced by aluminium. The sheets lie symmetrically one above the other;

---

[1] The electrical charges have been much studied by R. K. Schofield, and much information about them is to be found in his papers.

when the models are set up it is seen that hexagonal holes in the lattices are opposite each other. These become occupied by potassium ions, which are the only ones (except the ammonium ion) that can fit in, their radius being 1·33Å; and their positive charges balance the additional negative charges resulting from the replacement of some of the silicon by aluminium ions; they also serve to bond the layers together. These potassium ions cannot be replaced by others, they are thus more firmly held than the potassium ions in the exchangeable bases, though less firmly than the atoms forming part of the structure of the lattice by isomorphous replacement. The sandwiches form compact units which are built up into the well-known sheets.

A modification known as illite is of frequent occurrence in British soils. Fewer of the silicons are replaced by aluminium and therefore the increase in negative charge is less. The holes are less regular and the potassium ions fewer; the exchange capacity is less, usually something between 20 and 40 milli-equivalents per 100 grams. The forces holding adjacent sandwiches together are weaker: water molecules can to some extent force a way between them causing a certain amount of expansion, but when the water evaporates they come together again.

Further changes result when more of the potassium is lost and another mineral, montmorillonite, is formed. The bonding becomes weaker. The sandwiches themselves are solid enough, but they are more easily separated than in the case of illite: water molecules can more readily push in between them causing greater expansion of the particles; on drying they contract again as if they were held by an elastic band. They can easily separate completely, making the particles smaller and smaller, so bringing into prominence the physical properties associated with smallness of size and the consequent increase of surface. The arrangement of the atoms in the lattice is such that the effective negative charges are all on the outside of the sandwich; they can attract any cations including the basic groups in proteins. The base exchange capacity is thus several times higher than for illite and is of the order of 100 milli-equivalents per 100 grams. Soils from basic igneous rocks tend to contain montmorillonite or vermiculite while those from granite, sedimentary or metamorphic rocks tend to contain illite.[1]

The kaolin type minerals differ greatly from the preceding. The

[1] W. A. Mitchell, *J. Soil Sci.* (1955) 6, 94–8. He deals with Scottish soils but the statement is generally true for temperate climates.

lattice is compact and its units resistant to breaking up. There is little isomorphous replacement and therefore little resulting development of surface electric charges; these occur mainly at the edges of the units where the hydroxyl groups are exposed giving a negative charge which however is so small that the base exchange capacity is only about 5 milli-equivalents per hundred grams compared with 20 to 40 for illite and 100 for montmorillonite. The sheets are not readily separated; a hundred of them may be piled one on top of the other to make a small crystal, and water molecules do not so readily move in and out between them as in the case of the mica group. The particles tend to be larger than those of the illites and montmorillonites and therefore surface phenomena are less pronounced.

All of the above mineral types occur in British soils. The clay fraction of Broadbalk field, Rothamsted (Clay-with-flints) contains some 60 per cent of illite and 20 per cent of kaolin with smaller quantities of iron and aluminium oxides. The Kimmeridge clay (Cambridge district) also consists mainly of illite and kaolin, and the Lias clay is mainly illite. Montmorillonite occurs along with hydrated mica and kaolin in London clay and the chalk soil; and the somewhat similar vermiculite, also with mica and kaolin in Weald clay; Oxford clay sampled near Kidlington is mainly a non-expanded (i.e. non-hydrated) mica with some kaolin, as is also the clay of the Inferior Oolite soil, except that in this case the mica is hydrated, i.e. expanded.

In any description of the soil there is always the danger of over-simplification and this is particularly true in dealing with the clay minerals. Those described here are the commonest in British soils so far as investigations have gone, but they must be regarded as types or patterns and not as simple definite compounds like the simple salts of classical chemistry that can be represented by definite formulae. Isomorphous replacement results in a good deal of variation and it is not necessarily the same in successive layers. The replacing atoms differ in size from the aluminium or the silicon and the result is to break the symmetry of the layers so that they may not lie neatly stacked one on top of the other. The difference in atomic size must not be too great, however; only atoms with radius within the range 0·4 to 0·8 Å can be introduced into the lattice, otherwise it would become too distorted or even break down.

The properties of the clay depend not only on the lattice but also on the exchangeable bases. In British soils calcium usually heavily preponderates and our clays can be called calcium clays. But in arid regions the clay may contain much more sodium and then it has quite different properties. A sodium clay is also formed when land lies under the sea for some time, as in the occasional floods in the Eastern Counties or the newly won soil in the reclamation of the Zuyder Zee in Holland. All our agricultural and gardening methods, and all our varieties of plants and our food habits, have been evolved to fit a calcium clay; they are unsuitable for a sodium clay. It would be possible to devise effective cultivation and to breed usable plants that would grow, but it is far easier to convert the sodium clay into a calcium clay. This is done by adding suitable dressings of gypsum—calcium sulphate—the calcium of which displaces the sodium which thereupon forms sodium sulphate, a soluble salt readily washed away by the rain. This method has been very successfully used in Holland and in arid and semi-arid regions.

The clay minerals remain stable so long as the soil is neutral or only slightly acid or alkaline, which is the normal condition in Great Britain. But in conditions acid beyond a certain limit drastic changes may occur which will be dealt with later.

All the characteristic clay properties are greatly modified by calcium carbonate and soil organic matter, and they disappear when the clay is heated sufficiently strongly; it then loses all its free and combined water and becomes simply brick.

The minerals so far considered occur in relatively large quantities and may be regarded as the framework of the soil. Certain other minerals mostly occurring only in relatively small amounts are of special importance, chief among them are calcium carbonate, calcium phosphate, potassium compounds, iron and manganese oxides.

Calcium carbonate is quantitatively the most variable constituent in the soil: it may range from 70 per cent or more down to nothing. Of all the major mineral constituents of the soil it is the one most easily removed as it dissolves in water containing carbon dioxide in solution and is readily carried away in the drainage water: of this, indeed, it is almost invariably the chief mineral constituent. It is not all lost, however: the clay minerals can pick up some of the calcium and as already stated it is in fact the principal exchangeable base.

Some calcium carbonate is formed during the weathering of the rocks and some during the decomposition of plant material. But the limestone and chalk formations contain large quantities of shells and other remains of marine organisms deposited while the region was submerged under the seas of the earlier geological ages. Three great deposits occur in Great Britain.

The Carboniferous Limestone is the oldest and most massive, the thickness of the formation varying from 800 to 4,000 feet; it comes to the surface in the West Country and the North as a hard rock forming the inland cliffs of Cheddar, the Avon Gorge below Bristol, the cliffs around Castleton and Matlock in Derbyshire, Ingleborough and other famous places in Yorkshire; it contains some marvellous caves formed where an underground stream has dissolved out the calcium carbonate, frequently depositing it as fantastic stalactites and stalagmites. Sometimes the stream emerges into the open as a petrifying spring like the famous Dropping Well at Knaresborough where articles put into the water rapidly become coated with calcium carbonate thrown out of solution as the carbon dioxide escapes from the water into the air. Frequently, however, the soils overlying Carboniferous Limestone contain little or no calcium carbonate, either because it has been washed out under the rather high rainfall, or because the soil is not really derived from the rock, but has been transported from some other formation.

The Oolite limestone stretches from Gloucestershire to Yorkshire: like Carboniferous Limestone it is not pure calcium carbonate but contains sufficient silicate mineral and iron compounds to form a fairly good layer of residual soil as the calcium carbonate gets washed out.

The chalk of the Eastern and Southern counties is purer, especially the Upper Chalk as shown by its very white colour; many of the soils overlying it have had other origins but those that are simply residues left when the calcium carbonate had been washed away are very thin, and a great depth of chalk had to be dissolved to form them.

The remarkable thing is that even after millions of years exposure to the washing action of rain water these deposits of calcium carbonate should be so thick. The chalk is up to 1,750 feet in thickness: it contains remains of microscopic foraminifera that lived in the cretaceous sea, but most of it is the result of a chemical precipitation in circumstances not clearly understood. The conditions must have been fairly uniform over many millions of years, and one cannot help wondering where all the calcium and carbon dioxide came from, and why so little other material was present. There was some soluble silica, but by processes not yet fully understood, it separated out in the Upper Chalk as flint in layers of nodules or in bands or veins.

Calcium carbonate is extremely important in the soil because it determines the soil reaction which as we shall see later profoundly affects the life in the soil; in its absence the soil may become acid; it has other actions also.

Calcium sulphate or gypsum occurs in soils, but usually only in arid regions: being somewhat soluble in water it is apt to get washed out in humid conditions. Deposits of gypsum occur in this country, as for instance in the Triassic beds in Nottinghamshire, the Permian beds in Durham and elsewhere: those at Billingham in Durham play an important part in the manufacture of sulphate of ammonia. Calcium commonly constitutes some 80 to 90 per cent of the exchangeable bases in the clay minerals in British soil.

Magnesium is vitally important to plant life: it occurs in some of the clay minerals and frequently comes second in abundance to calcium among the exchangeable bases. It may also occur as carbonate in the soil, and, being less soluble than calcium carbonate, is less liable to be washed out; large quantities occur in union with calcium carbonate in the magnesian limestones of the north and west of England.

Potassium has presented many difficult problems to soil chemists. It occurs in certain silicate minerals, in some it is in such resistant combination as to be useless for plant food; in others, orthoclase and some of the mica group, especially illite where it fits into pockets in the lattice, it is rather more available; but it is most available as an exchangeable base, and this is usually its most important form for plants. The exchangeable potassium, however, forms only about 1 per cent of the total potassium in the soil, but fortunately it is renewable: it appears to be in equilibrium with the potassium in the lattice and consequently the quantity taken up by the plant is replaced. W. E. Chambers at Rothamsted found that wheat grown year after year without potassium fertilizer on the Broadbalk field had between 1843 and 1921 taken 12 cwt. per acre of potassium from the soil, but the exchangeable potassium fell by only about 1·6 cwt. per acre. On the adjoining plot receiving potassium fertilizer nearly 60 cwt. per acre of potassium had been supplied, about 28 cwt. had been removed in the crop, some—probably not much—was washed out in the drainage water, but of the 32 cwt. not accounted for only $2\frac{1}{2}$ cwt. had been added to the exchangeable potassium. The remainder had become insoluble and non-exchangeable, nevertheless it can replace the exchangeable potassium that has been taken by the plant.

Iron oxide—ferric oxide—is one of the most conspicuous constituents of the soil. It is usually hydrated and it can be either red or yellow; it occurs in two forms: as a jelly coating the soil particles like paint and giving them their distinctive colours modified by brown or black humus; and as fine amorphous particles in the clay fraction. Colour is no guide to the amount of iron oxide present: a white chalk soil may contain 3 per cent, the same as in a reddish clay soil; even the bright red soils of Devon do not contain more than some of those of quieter colour. Like the clay minerals it possesses colloidal properties: it has, however, no power of base exchange but in acid conditions it can bind certain anions, i.e. acidic radicles, notably phosphate, and it then interferes greatly with the phosphate nutrition of plants. It is insoluble in water but it has the remarkable property of entering readily into soluble or at least mobile combinations with certain organic compounds: water leaching through leaves picks up something that dissolves it and it can then move downwards in the soil especially if the soil is sufficiently acid. This action is very important in connection with the formation of podzols (p. 195). The soil particles readily "sorb" some of it, i.e. become coated, and a relatively small change in conditions may throw it out of solution; once this process starts it continues. So it often happens on acid sandy soils that the iron is washed out of the surface 6 or 8 inches and precipitated as a band in the next few inches below; this is called an iron pan and is of special importance wherever it occurs.

Ferric oxide is fully oxidised and is not only harmless to plants but is an essential nutrient. In absence of air, however, it is liable to lose its red colour and become reduced to the ferrous state which is harmful to plants: this occurs when soil remains water logged for a sufficient time. If, as often happens, a sulphate is present, this also may be reduced forming black ferrous sulphide from which in acid conditions poisonous sulphuretted hydrogen is evolved. When the water is drained away and air admitted the ferrous iron is oxidised to ferric and then becomes harmless. Certain micro-organisms play an important part in both oxidation and reduction: this latter can also be brought about by some of the plant degradation products.

The alternate reduction and oxidation brought about by movements up and down of the water table produces certain typical effects on the zone affected, notably mottled rust-coloured patches due to changes in the iron compounds: this condition is called by its Russian name "gley".

Aluminium oxide is closely associated with iron oxide and like it, occurs as a jelly coating the soil particles and as minute amorphous particles in the clay fraction, but being colourless is not seen in ordinary soils. Like ferric oxide it is insoluble in pure water but dissolves in water that has washed through certain leaves and so contains the appropriate organic compounds. It can thus become mobile in the soil, but it can be precipitated in conditions similar to those for iron. It has only one form of oxidation and is therefore not liable to reduction, but in its soluble form in acid soils it is poisonous to many plants; it can however be thrown out of solution and rendered harmless by addition of lime. Plants vary in their susceptibility to it: some—a rather large number in fact—not only tolerate it but accumulate some of it in their tissues.

All three manganese oxides may occur in soil: the tetravalent (MnO_2) is found as black nodules in some of the clays, e.g. the Clay-with-Flints in Hertfordshire; the others occur as hydrated forms. As with iron, the higher oxides may be reduced to lower, and the lower oxidised to higher forms according to the conditions; both changes can be brought about by micro organisms.

The phosphorus compounds in the soil are derived originally from the rocks but much of the phosphate has already been taken up by plants, built up into their tissues and returned to the soil in organic combination when the plant dies and its residues become incorporated with the soil. In the sedimentary soils some of the phosphate has formed part of the sea animals the bodies of which got caught up in the deposited material.

Most of the inorganic phosphate is an apatite, $Ca(OH)_2, 3Ca_3(PO_4)_2$ or a fluorapatite in which fluorine replaces the OH groups. Immense deposits occur in certain places: North Africa, Tennessee, Carolina, parts of the U.S.S.R.; it is curious that they should be so very localised and unfortunate that the British Commonwealth possess none of them. Apatite is insoluble in pure water, but it is slightly soluble in soil water as this always contains carbonic acid, and a certain amount of movement then becomes possible especially if the soil is acid. But the phosphate is so readily absorbed by some of the soil colloids, and taken up by iron and aluminium oxides, that only traces of it ever appear in the drainage water. Only about half the phosphorus in British soils is in inorganic form, however; the rest is organic and is described later.

Sulphur occurs in practically all soils. It came originally from the rocks in which it occurs as sulphide, but this is rapidly oxidised to sulphate which however in waterlogged conditions may be reduced to sulphide again: micro-organisms play an important part. Like phosphorus it is an integral constituent of plants, and is returned to the soil in organic form when the plant dies.

Soil always contains small quantities of soluble salts; these are of vital importance as plant nutrients and will be discussed in detail later. A rough idea of their composition can be obtained by analysing shallow well water, or better, water discharged from field drains. In Great Britain and North America the chief base is calcium followed a long way behind by hydrogen, sodium, magnesium, potassium, iron, aluminium and much smaller quantities of other metals; the chief acid radicles are carbonic, nitric, sulphuric, hydrochloric and silicic. It is not far wrong to think of the dissolved salts as a mixture of calcium bicarbonate, calcium nitrate and smaller quantities of others. The water is "hard" and refuses to form a proper lather with soap: also when boiled it deposits a "fur" of calcium carbonate on the side of the kettle or saucepan. The drainage water only incompletely represents the soluble salts, however; absorption and releases by the soil colloids, and exchanges with the clay minerals, make any precise estimate of their nature and amount impossible.

We are now in a position to form a mental picture of the mineral constitution of the soil. The skeleton is formed of particles of highly resistant inert and biologically useless quartz or unchanged rock. These are coated with a jelly consisting of iron and aluminium oxides, very finely divided clay minerals, and some organic matter which, as will be shown later, is an important part of the habitat of some of the soil population. (Plate III, p. 66). The finest fraction of the soil particles, the clay, consists of finely powdered quartz and other stable materials, of amorphous oxides of iron, aluminium and manganese in small amounts, and of minute crystals of silicates derived from the

b. Calcium deficiency: tomatoes.

a. Iron deficiency: sweet cherry leaves.

Plate 1. Leaf symptons resulting from deficiencies of certain mineral nutrients (*by permission of the Controller of H.M. Stationary Office*)

original rock by weathering. These can be put into two groups, one of which, the micaceous, is chemically active by reason of the electric charges on its constituent atoms; those on the surface of the lattices are negative and therefore attract the positively charged basic cations in the soil moisture: these can be displaced by other cations. The basic elements that will frequently be mentioned in this book because of their dominating influence on the soil population: calcium, magnesium, potassium, iron, aluminium and manganese, occur in several forms in the soil. All except potassium may occur as isomorphous replacers in the lattice of the clay minerals in which form they are so stable as to be practically inert; potassium may be held in a pocket in the lattice, from which it is dislodged only with difficulty; calcium, magnesium and potassium may be held as exchangeable bases which can readily be picked up by plants and other organisms; any can be held by "sorption" on the surface of the soil colloids; some can occur as simple compounds: calcium and magnesium as carbonates, iron, aluminium and manganese as oxides in the form of jellies, minute particles or nodules. The acidic anions, phosphate and sulphate, exist in organic and inorganic form. The phosphate is only to a very slight extent washed out from the soil and so is used over and over again by plants and soil organisms. In English soils some 30 to 70 per cent of it is in organic form but it is continuously being changed back to the mineral or inorganic form by certain micro-organisms and then taken up by a new generation of plants and once more built up into organic combination so going through an endless cycle: so with sulphur.

This picture omits many details but it is not over-simplified to the stage of distortion: it will serve to warn the reader against any of the easy generalisations that come of incomplete knowledge. The constitution of the soil is complex and is under investigation in many laboratories in different countries: more is discovered about it every month, the foregoing account is in general accordance with present day knowledge.

NOTE TO CHAPTER I

J. G. D. Clarke in his *Prehistoric Europe*, (1952) sets out the climatic and vegetation history of post-glacial Denmark (which would be typical of N.W. Europe generally) as follows:

LATE GLACIAL: Sub-arctic tundra, with warm oscillation *c.* 10,000–9,000 B.C. marked by birch-forest.

PRE-BOREAL: Slow rise of temperature, beginning *c.* 8000 B.C.

BOREAL: Rising temperature, continental climate, beginning *c.*6800 B.C. Pine/birch forest followed by Pine/hazel, and beginning of mixed oak forest.

ATLANTIC: Warm, moist (i.e. oceanic), beginning *c.* 5000 B.C. Mixed Oak-forest (Oak, elm, lime) and alder.

SUB-BOREAL: Drier, more continental, beginning *c.* 2500 B.C. (I) Introduction of cereals and weeds of cultivation; (II) Oak forest; (III) Spread of grasses and ling.

SUB-ATLANTIC: Colder and wetter, beginning *c.* 500 B.C. Beech, Pine revertence.

A full account of the geological actions that have moulded our landscape is given by L. Dudley Stamp in *Britain's Structure and Scenery*. The secular variations of our climate are discussed in detail by Gordon Manley in *Climate and the British Scene*. Both volumes are in this series. See also H. Godwin, Presidential Address to Section K, *The Advancement of Science, 1956*.

CHAPTER 2

COMPLETING THE SOIL: THE
ORGANIC MATTER

THE SYSTEM of mineral particles and compounds described in the last chapter, complex though it is, could support neither plant nor animal life; it lacks the nitrogen compounds essential for plant growth, and the protein and energy-giving compounds indispensible for animal life. There is no lack of free nitrogen in the atmosphere, but neither plants nor animals can use it; only when it has been converted into ammonia, or, better, nitrate, does it become available as plant food. As is shown later certain micro-organisms have the power of assimilating gaseous nitrogen and converting it into protein which others convert into ammonia and nitrate: it may be assumed that these nitrogen fixers were among the earliest living things on the earth and that they had to build up a stock of fixed nitrogen before the first higher plants could begin to appear.

All living things require sources of energy to operate their vital processes. Green plants are unique in that they can utilise the energy of sunlight, and with this they can synthesise sugars from which they can build up cellulose, lignin and other structure material as well as oils, and, with nitrate, phosphate and other substances taken up by their roots they synthesise nucleoproteins, proteins and other compounds in all of which some of this energy is stored. When the plants die their roots and parts of their leaves and stems mingle with the soil and add to it a new factor: a source of food and energy indispensable to all living things except green plants and a few lowly forms of life that make their own provision. These sources, with the air, water and mineral substances already present in the soil make it suitable for life, and it is a general rule that wherever life can appear it does. A highly complex population of organisms adapted to the conditions in the soil has evolved which, in living on the organic matter added by the plants, transforms some of it into substances of the greatest importance to the soil itself, including also essential foods for later generations of plants. While the first plant colonisers derived their mineral food only from the weathered rock materials those coming later derived much of it from the decomposition of the residues of their predecessors.

It would be possible to reconstruct the process by watching the colonisation on a ton or so of a deep subsoil in a concrete bin kept in the open air and provided with holes at its base to allow free water to drain away. The more effective way is to investigate in the laboratory the course of decomposition of plant materials as it might be expected to occur in the soil, and to study chemically the organic matter of the soil to see how far its composition accords with the conclusions so reached.

Most of the organic matter of plants belongs to one or other of three groups of substances: carbohydrates, lignins, and proteins. The carbohydrates are the largest in amount and the most varied, ranging from easily decomposable sugars to the polysaccharides and the more resistant celluloses. By the time the plant is dead the sugars are probably gone but the polysaccharides and the celluloses remain; they, however, are attacked by micro-organisms: part is broken down to carbon dioxide and water, part is built up into the body tissues of the organisms, and part is synthesised into more complex substances, amino-sugars, mucopolysaccharides, and others. The lignin is attacked much more slowly and incompletely: it undergoes a certain amount of hydrolysis and oxidation; some of its hydroxyl and methoxyl groups are converted into carboxyl groups, and it associates with protein to form a black complex. Protein by itself is rapidly decomposed in the soil; most of it goes to carbon dioxide, water and ammonia which is speedily oxidised to nitrate: there is some building up of complex compounds by the soil micro-organisms, but vegetable protein incorporated in the soil may have some 80 per cent of its nitrogen converted to its end product, nitrate. Protein in association with partly decomposed and oxidised lignin, however, is much more resistant; its nitrogen is very firmly held, and is only slowly converted into nitrate.

These decomposition products, together with dead soil organisms, fungus mycelium, and finely divided plant residues which have lost their structure (but not visible fragments which retain their structure, and which are always excluded) constitute the true soil organic matter. In many ways it behaves as if it were a single substance (which it is not) and it is generally called humus: it is black, structureless, and amorphous, in the form of very minute particles or a jelly coating the sand, silt and clay particles. It has strongly developed colloid properties: it swells on wetting, shrinks on drying, has some power of sorbing ions; it can also hold and exchange bases. In all these properties it resembles the mica group of the clay minerals but with this fundamental difference: the humus contains true acid groups by which the bases are held, while the clay does not. Humus and clay particles can hold each other firmly, and the higher the charge on the clay the more stable the combination. There are probably some positively charged groups on acid humus but usually the negative groups preponderate.

In spite of a prodigous amount of work little is known of the chemical structure of humus: as it does not form crystals neither X ray photography nor the electron microscope has revealed much, and the standard chemical methods have given little or no information about the nucleus, although they have given some about the side chains. A considerable part—40 to 60 per cent—of the humus is an acid called humic acid combined in normal soils with bases, predominantly calcium: these can be removed with dilute hydrochloric acid. The acid part, which is insoluble in water, can be extracted with dilute alkali with which it forms a dark brown solution; from this it can be precipitated on addition of acid as a black sticky mass; it dries, like clay to horn-like flakes. As prepared at Rothamsted it contained 56 per cent of carbon,[1] 5·3 per cent of nitrogen and 5·1 per cent of hydrogen. The soil still remains black, however; the considerable organic residue still left is called humin. It has no marked chemical or biological properties, and may be chiefly a condensation product of humic acid but it also includes some finely divided carbon.

As so little is known about the constituents of the soil organic matter it is best considered as a whole. Its nitrogen is especially important as a source of food for soil organisms and plants. About one third to one half of it is in the form of protein, and J. M. Bremner of Rothamsted, who has studied this subject extensively, has shown that the protein is generally similar in chemical constitution in all the soils examined. About 5 to 10 per cent of the nitrogen is in the form of amino-sugars and about 40 to 50 per cent is as yet unidentified. Phosphorus also is present: usually indeed more than half the total phosphorus of surface soils is organic. In the Scottish soils examined

[1] Aslyng (*Acta Agric. Scand.* (1956), 6. p. 64, gives 56 to 58 per cent as a usual range.

by E. G. Williams and W. H. H. Saunders[1] the range was 27 to 67 per cent and it occurred mainly in the clay and silt fractions in contradistinction to the inorganic phosphorus, which occurred chiefly in the fine sand. The chief organic compounds are inositol phosphate, smaller quantities of phytin, the hexaphosphate and the nucleic acids and the decomposition products. Recent work at the Iowa Experiment Station has however thrown some doubt on some of the identifications.[2] Whatever the nature of the organic phosphorus compounds some are decomposed by soil organisms and become available as plant food.

Proteins and amino-compounds would by themselves be rapidly decomposed by the soil organisms, but the clay minerals appear to hold them so firmly, as also does the altered lignin, that the soil organisms cannot readily dislodge the nitrogen group to oxidise it. The clay minerals similarly protect the organic phosphorus compounds against attack.

This protection of the nitrogen group against rapid attack by microorganisms has been one of the most important factors in the development of our landscape and scenery. For without a continuous supply of nitrates in the soil few of the higher plants—trees, shrubs, flowering plants and grasses—could grow. Protein from dead plant material is the chief natural source; by itself as we have already seen it is rapidly decomposed and the nitrogen appears as ammonia and then nitrate. But nitrates cannot be sorted for long in the soil: of all plant foods they are the most easily washed out by rain. The protection of proteins by the clay colloids and the lignin slows down the decomposition to such good effect that on balance only about 1 per cent of the total soil nitrogen becomes available in any one growing season, thus assuring a continuous supply even though it may be only on a low level.

The carbon is less protected than the nitrogen, and in consequence the ratio of carbon to nitrogen which in the dead plant material would range about 25 to 40[3] is generally reduced in the soil organic matter to about 10 in arable soils and rather higher in pasture soils. Ratios of

[1] *J. Soil Sci.* (1956) 7, 90-108.

[2] Adams, A. P. *et al. Soil Sci. Soc. Amer. Proc.* (1954) 18, 40-46.

[3] In the roots and stubbles of cereal crops at Rothamsted the ratio is about 43, of leguminous crops 23 and of moderately rotted farm manure 18.

this order are found in a wide range of soils provided they are not too acid, showing that some sort of equilibrium is attained which is not very different whether the amount of organic matter added to the soil is rather high or moderately low. So also the quantity of organic matter remaining in the soil reaches an equilibrium depending on the rate at which fresh organic matter is added and the rate at which the soil organisms can decompose it. Provided the conditions are favourable for the organisms their numbers and total activity increase as the supply of organic matter increases. But there is an upper limit, set by the conditions, beyond which the organisms cannot go; the organic matter then accumulates only partially decomposed and the soil loses its character; if this results from a too rapid feeding in of organic matter the soil simply becomes a compost heap; if it comes about through a worsening of conditions, e.g. exclusion of air, or increasing acidity, the soil becomes a peat moor or bog.

The level of organic matter finally attained in the soil is greater under grass than under trees because the grass roots are finer, more easily decomposed and tend to concentrate in the top 9—12 inches of soil, though some roots go deeper. The level falls on cultivation because the increased air supply stimulates the soil populations to greater activity in decomposing the soil organic matter; it is raised by increasing the water supply in the soil because this increases the growth of roots and leaves and does not correspondingly increase the activity of the soil organisms; it is raised also as the temperature falls because this slows down the rate of decomposition more than the growth of the plant.

The direct way of following these changes would be to determine the percentage of organic matter in the soil, but the simpler and more useful plan is to determine the percentage of nitrogen, the fluctuations in which are parallel with those of the organic matter. The highest percentages of nitrogen are formed in some of the old pastures, especially the riverside marshes sufficiently provided with calcium carbonate where the ample water supply favours abundant growth of grass, but slows down decomposition. In some of the Thames marshes the percentage of nitrogen rises to 0·6;[1] this value recurs in certain very rich pastures: it corresponds to about 12 per cent of organic matter in the soil and in the writer's experience is the highest reached in Britain

---

[1] In all these cases the values refer to a depth of 9 inches. Similar figures are obtained for North American soils, and there also the percentage of organic matter is roughly 20 times the percentage of nitrogen.

in true soils other than peat or fen. Dryland rich pasture soils in Romney Marsh, the Gault clay in Kent and elsewhere have a usual limit of about 0·3 to 0·4 per cent of nitrogen, about 6 to 8 per cent of organic matter; for the rich unbroken Canadian prairies the value is about the same. Old pasture on poorer soil contains less organic matter: at Rothamsted the limit is about 0·25 per cent of nitrogen, about 5 or 6 per cent of organic matter.

But when the pasture is broken up for arable cropping or conversion into gardens the organic matter rapidly lessens in amount. After 22 years of cultivation the nitrogen in a Canadian prairie soil had fallen from 0·371 per cent to 0·254 per cent, a loss of 2,190 lb. per acre, only a third of which had been recovered in the crop. This was not the result of bad management but the inevitable consequence of changing from a perpetual vegetation cover to cultivation conditions: it is doubtful whether any system of management of wheatlands could have kept the nitrogen content at its original level.

When old grass land at Rothamsted is broken up its organic matter falls in amount and if none is added as manure the nitrogen content ultimately drops from about 0·25 per cent to just below 0·1 per cent.— about 2 per cent of humus—but falls no lower, however long the land is left without manure. This represents the lower limit not only at Rothamsted but also at Woburn on a lighter soil and in the extensive series of soils studied by A. D. Hall and the author in their survey of Kent, Surrey and Sussex. The amount of organic matter in a cultivated soil can be raised by working in compost or farmyard manure though hardly to the level reached in an old meadow. The Broadbalk wheat field at Rothamsted has been in arable cultivation as far back as records exist: it has been cropped annually with wheat since 1843. One of its plots has had no manure of any sort since 1839, an adjoining one has had 14 tons per acre of farmyard manure annually since 1843. The percentages of nitrogen in the soils of the two plots have been:—

	1843	1865	1881	1893	1936	1945
No manure since 1839		0·105	0·101	0·094	0·103	0·105
14 tons f.y.m. annually	0·12[1]					
since 1843		0·175	0·184	0·213	0·226	0·236

[1] Sir Henry Gilbert's estimate. All the Rothamsted soil data refer to the top 9 inches of soil. Other analysis by, or checked by R. G. Warren.

The slight rise on the unmanured plot since 1893 may be due to improved methods of cultivation after the 1920's which resulted in higher crops and therefore larger root and stubble residues. This factor also operated on the farmyard manure plot and may account for some at least of the rise. The unmanured plot reached its steady state in about 20 years while the manured plot took longer: it rapidly gained organic matter and nitrogen in the first twenty years, then the increase slowed down and is now detectable only by careful analytical procedure. Even after a hundred years of annual dressings of farmyard manure—by which time the soil had received 1,400 tons per acre, its own dry weight down to a depth of nine inches—its nitrogen content was only 0·236 per cent, and had not yet reached the 0·25 per cent of the old grass land at Rothamsted.

Practically the whole of the manure added during the latter years thus appears to have been dissipated into the air and drainage water leaving only little trace in the soil. This however is not quite correct: some of it has made good the loss of organic matter and nitrogen that the soil continuously suffers. It does however seem to be true that the later dressings are decomposed more completely than were the earlier ones when the soil was still well below its upper limit of organic matter content. The population of organisms is now much larger, and the clay minerals are presumably more fully saturated with organic matter and therefore less able to protect from attack the nitrogen groups added later.

In grassland the upper limit of organic matter and nitrogen content is maintained indefinitely so long as the conditions are not changed because the grass being a permanent crop is continuously shedding more organic matter into the soil. But on cultivated land the upper limit is in a much more precarious condition and is maintained only by continued additions of manure. If this is withheld the percentage of organic matter falls, rapidly at first and then more slowly. On the Hoos barley field at Rothamsted an experiment parallel to that on the wheat field was started in 1852. One plot was kept continuously without manure while the adjoining plot received annually 10 tons of farmyard manure per acre. Barley was then grown every year on both. In 1871 the dunged plot was divided into two: one part continued to receive its manure as usual, the other was kept unmanured; the cropping was continued as before. In 1946 the percentage of nitrogen in the continuously unmanured soil was down to the Rothamsted minimum of 0·1 per cent and no doubt had been for some time; in the soil receiving farmyard manure every year it was up to the Rothamsted maximum; but on the soil that had received the annual dressing of manure between 1852 and 1871 and then no more the nitrogen was still well above that for the unmanured plot. Even 75 years of cropping without manure had not sufficed to remove all the nitrogen added in the 20 annual dressings of farmyard manure previously applied:—

TREATMENT	*No manure since 1852*	*Farmyard manure 14 tons annually since 1852*	*Farmyard manure 14 tons annually 1852–1871 then unmanured*
PER CENT OF NITROGEN IN 1946	0·103 (*Rothamsted minimum*)	0·272 (*Rothamsted maximum*)	0·151 (*R. G. Warren*)

The crop yields differ in the same direction.

This slow rate of change is sometimes taken to mean that plant residues and farmyard manure contain some substances so stable that they persist for long periods in the soil suffering regularly a small amount of decomposition each year. It is known that the complexes formed with lignin and the montmorillonite group of clay minerals are very stable yet it is difficult to see why the decomposition should be so long drawn out. On the other hand the cycles of biological and chemical changes in the soil explain the observed facts very well. The organisms decompose the added organic matter, but they also build up some new organic compounds in doing so; decomposition and synthesis thus go on side by side but certain products are continuously thrown out of the cycle, notably carbon dioxide, water and nitrates. These account for the losses; the simultaneous synthetic processes explain the slowness with which they take place and the long survival of the organic matter in the soil. The classical chemical methods can record only the net changes and hitherto it has been impossible to estimate the magnitude of the decomposition and the synthetic processes. The isotopes now made available at Harwell will, however, make it possible in the near future to follow the changes in much more detail.

A clod of earth seems at first sight to be the embodiment of the stillness of death, but its apparent quiescence is completely illusory; physical, chemical and biological processes are ceaselessly active, bringing about continuous cycles of change, some upgrading, some downgrading, but buffered and saved from violence by the clay and organic matter. A steady balance is thus maintained so long as the soil conditions do not greatly alter, but it shifts one way or the other when conditions change. The study of these shiftings and changes has revealed some extraordinarily interesting results which will be set out in the following chapters.

THE ARCHITECTURE OF THE SOIL: SOIL STRUCTURE[1]

The mineral and organic substances described in the preceding pages do not form a simple mixture in the soil: excepting on very sandy soils they tend to build up into crumbs of various sizes up to about one eighth of an inch in diameter each of which is almost a complete miniature of the soil itself. They can be seen by crushing in the hand a small dried clod of good garden soil: it breaks down quite easily to crumbs which do not as readily break down further. This crumb structure is specially marked in grassland: the grass roots do something which causes the soil particles to unite. Grasses are not all equally effective; those having a considerable quantity of strong roots produce more crumbs than those with smaller or more fibrous roots; clover grown by itself is not so good as the effective grasses though it may enhance the grass action when both are grown together. Lucerne is effective, though not usually as good as grass.

The difference in soil structure under grass and on adjoining cultivated land is obvious on mere inspection. On the Rothamsted soil this does not appear to result from a greater water stability of the crumbs under grass but from less exposure to water with the result that they persist longer than on cultivated land which in wet weather tends to become water-logged. Dead grass roots can suck water from the surface into the lower level of the soil thus saving the crumbs from dispersing. The length of time needed to produce a significant effect depends on the conditions: on the Lower Lias Clay at the Drayton Grassland Improvement Station near Stratford on Avon crumb formation was very marked after four years under a good ley; it was greatly diminished after four years in arable cultivation. On the less tenacious clay at Rothamsted ten years under grass hardly sufficed to

[1] Soil scientists use the word "texture" in reference to the size of the soil particles, so distinguishing sandy soils, loams, clay soils, etc, and "structure" for the way in which the particles are built up, as described in this section.

effect much change though a difference was seen after 13 years. But whether a longer or shorter time is necessary the crumbs form.

They are of varying degrees of stability. Some can stand up to gently falling drops of water, as in the usual rainy day in Britain, while others simply collapse and are reduced to particles so fine that they coat the surface of the soil with a layer that glazes it: the former are called "water stable", the latter "water unstable". A soil with stable crumb structure can safely be cultivated over a fairly wide range of moisture content, but where the crumbs are unstable the safety range is much narrower, and the crumbs are easily broken down if the soil is worked when too wet. The glazing of the surface retards the soaking of the rainwater into the soil, and the break down of the crumbs in the body of the soil puts it into a poached and sticky condition unfavourable to the plant roots.

The building up into crumbs has several important effects on the soil. The finest particles are held in the crumb and are no longer free to block up minute passages in the soil down which the rain water seeps to the lower depths. A soil with good crumb structure is therefore well drained provided of course that there is no hindrance to the subsequent movement of the water. The crumbs being irregular in shape do not completely fill the space they occupy: much is left empty so that air can get in and out easily. The crumbs themselves also are porous and allow this two way movement of air. And although a good crumbly soil may dry into clods these are very different from those formed where the crumbs are unstable. Clods of good crumbly soil can fairly readily be worked down to the mellow tilth so necessary for a successful seed bed, while those formed from unstable crumbs are hard when dry and very sticky when wet; only over a narrow range of moisture content can they be made to produce a tilth and then not always a good one.

A great amount of work has been done in recent years to discover how crumbs are formed in the soil: the story is still far from complete but much has already been learned and the way is now open to learning much more. They may arise either from the breaking down of clods or from the building up of the individual particles. Breaking down is done artificially by suitable cultivation processes which will be described later; by far the most effective natural agent is frost. The building up necessitates a mixing agent as the crumb contains all the soil constituents, usually also some of the organic matter that has been added in the form of plant residues or farmyard or other manure.

Cultivation brings about a good deal of mixing but the most effective natural agents are some of the earthworms as will be shown later. Earthworms have sometimes been credited with actual crumb formation but this does not appear to be the case except where organic matter is present: R. J. Swaby at Rothamsted showed that worm casts from old grassland were more water stable than aggregates from the grassland itself, but on arable land there was no difference.

Clays differ in the stability of the crumbs they form. Crumbs from sodium clays are stable only in presence of a certain amount of sodium chloride or other sodium salt; interesting instances occur on land that has been flooded with sea water in the eastern counties of England. After the water has drained away the soils work easily for a time, but when rain has washed out the sea salt they become sticky and more difficult to cultivate. Crumbs formed by acid clays are even more stable than those formed by calcium clays, the normal clays in Britain.

It is not known how the various particles come to unite to form the crumb. Inter-particle bonds may develop analogous to weak residual affinity of chemical nature, a view that is held by some of the American workers, or one may simply say that the particles are stuck together by some cementing material. The crumbs can build up only to a certain size; beyond that they become unstable and break down easily.

Several glue-like or colloidal substances capable of acting as cements exist in the soil, particularly the red or yellow ferric hydroxide, the white aluminium hydroxide, and the black humus. The uniformity of colouring of the soil suggests that they coat the particles of sand and silt: the electron microscope shows that such a coating does in fact exist; (Plate III, p. 66) it is confined to the particle, however: there is no evidence of a capsule round the crumb or of flocculated clay enmeshing it.

The parallelism between crumb formation and accumulation of organic matter in the soil suggests that the two processes are closely connected. Both are at their highest development on old grassland where dead roots and leaves are continuously being mingled with the soil; both suffer when the grassland is broken up, and are at their lowest condition in soils long cultivated without addition of organic manure or grass and clover crops; partial restoration for both can be effected by periodical and substantial dressings of farmyard manure.

This close relationship is largely due to the soil micro-organisms. Their direct action does not seem to be very marked: fungi and to a less extent myxomycetes have some small temporary effect, the fungus mycelium serving as a binding agent until it is decomposed by bacteria. But some of the products of micro-organic decomposition of the plant residues are very effective crumb formers, especially apparently some of the synthetic products. Like the fungus mycelium, however, they do not long persist in the soil but are decomposed by bacteria; fresh supplies are constantly required hence the necessity of frequent additions of plant material to the soil and the advantage of grassland, where addition is continuous, over cultivated land where it is only occasional.

It is not known what the effective compounds are in the soil, but a number of the multitude of possible substances have been investigated in the laboratory. Two types of action have been discovered. Some of the organic substances coat the clay particles so modifying their properties, reducing for example the swelling of the clay on wetting, also the stickiness and therefore the tendency to form dense clods. Others act as binding agents uniting the particles one with the other. The electron microscope shows that these are uniformly distributed throughout the crumb and in contact with each particle. The most effective are polymers having chains of carbon atoms long enough to reach from particle to particle, capable of being adsorbed at anchor points on the clay, well provided with the active groups $-OH$, $-NH_2$, and $-COOH$ and not cluttered up with side chains in such a way as to make their shape unsuitable. Their molecules are very large: one of the most active is said to have a molecular weight of about 4 millions.[1] This appears to be exceptional: it is recognised however that the molecular weight should be of the order of 50,000 to 100,000 or more. Cross linked polymers are not as effective as linear polymers. Compounds of this kind, polysaccharides of the polyuronide type, are known to exist in the soil and they may play an important part in crumb cementing and stabilisation.

Whatever the natural agents may be they are readily decomposed by soil micro-organisms. Modern chemical technique can, however, synthesize polymers of long chain type that also produce crumbs but are not nearly so easily decomposed and remain stable in the soil for

[1] Dextran from the Leuconostoc group of organisms (R. L. Whistler & C. L. Smart, *Polysaccharide Chemistry*, Academic Press, N.Y. 1953).

long periods. Some of these have been studied at the I.C.I. Jealott's Hill Research Station, Bracknell by M. J. R. Geoghegan and others; while the Monsanto Chemical Company has prepared some of them on the large scale and put them on the market under the name Krilium. The patent describes them as "ethylenic polymers having numerous side chains distributed along a . . . linear carbon atom molecule" and it covers 61 products.

Krilium has proved useful in preventing the formation of surface crusts and in producing a good crumb structure on heavy soils and so securing better germination and an earlier start for seedlings, but it has not in Great Britain brought about any increase in yield in field or garden trials. The effect may not have lasted long enough, or a good crumb structure may be less important during the growing period than is usually believed. The important point, however, is not whether a particular product is or is not useful in agriculture: it is that a way has been opened up for controlling crumb formation in soil, and once that has happened there is no knowing where it may lead.

THE PORE SPACE

The crumbs being formed of particles of varying shapes and sizes are not solid throughout but are porous, like miniature sponges. They also are irregular in shape and do not fit closely together like bricks; spaces are left between them. The individual spaces are so small that they cannot be seen by the naked eye, but they are so numerous that the aggregate pore space in and between the crumbs is almost unbelievably large. A clod of earth may look absolutely solid throughout yet more than half of it may be empty space—empty except for air and water. On soils of the same type those in arable cultivation and un-manured have the smallest amount of pore space, and those in old meadows the largest: the range may be from 40 to 60 per cent or more. Well manured and cultivated soils come in between, standing higher in the scale according as the manuring has been long continued.

The spaces are so irregular in shape that no statement of individual dimensions would convey any meaning. Instead they are specified by the suction needed to empty them of water, which can be accurately measured. This value is then translated into terms of size by expressing it as the diameter of a circular tube which, if filled with water, would require the same suction to empty it; this is dealt with more fully later (p. 40).

These surfaces and pore spaces are of vital importance to the inhabitants of the soil, for it is there that they live their lives and from them draw the air, water, and much of the food on which they are dependent.

THE SURFACE AND THE SUBSOIL

Anyone acquiring a garden should dig down to a depth of two or three feet if he can to see what lies below the surface. In general in Great Britain he will find a fairly well marked distinction between the top layer, which may be 6 to 12 inches in depth, and the lower layer. The top is darker in colour, less sticky, looser, usually crumb-like in structure in an established garden, though not always in a new garden if it is on a stiff clay. The reason for the difference from the subsoil is that the organic matter which is necessary to complete the soil and which gives the dark colour has come from the soft roots of grasses and herbaceous plants and from plant residues falling on the surface and carried down into the soil by worms and other animals, or mixed in by digging, and none of these causes usually operates at more than about 6 to 12 inches below the surface.

The difference in the amount of organic matter accounts for part of the difference in structure and in stickiness but several other causes operate also. Heavy rain beating on the surface of the soil shatters some of the crumbs and washes some of the finest particles into the subsoil, making it heavier and more sticky. The top layer is subject to much greater variations in temperature and water content than the lower: wetting and drying, freezing and thawing, are more frequent, and this conduces to crumb formation. The surface soil contains more plant food constituents than the subsoil, some derived from the greater amount of plant residues present, some from accumulations of previous manurial dressings; some of it has been brought up from the lower depths by tree roots, translocated to the leaves which fell on to the surface of the soil and decomposed, liberating the plant food once more. The more open structure of the surface soil makes it a better habitat for plants and organisms; these are much more numerous than in the subsoil and bring about changes described later. On some of the heavy loams there are cracks in the subsoil down which water can soak away, thus saving the surface soil from becoming water logged after heavy rain (Plate XVI, p. 207).

It is sometimes thought that the subsoil represents the virgin soil and is therefore richer in plant food. This is a great mistake. The subsoil may even contain harmful substances resulting from the reduction of sulphates or of ferric or manganic oxides.

The above description applies to the Brown Earths to which great areas of British soils belong. Under high rainfall, or where the soils are very porous, some podzolisation occurs (p. 16) causing certain modifications.

CHAPTER 3

THE LIVING CONDITIONS FOR THE
SOIL INHABITANTS

A LL LIVING things have certain requirements in common. All need oxygen: the vast majority can take it only from the air, though a few very lowly forms of life can obtain it from certain of its compounds, especially nitrates, sulphates, and the higher oxides of iron and manganese. All require water in relatively large amounts, and all function only within a certain limited range of temperature. Each needs its own type of food. Most soil inhabitants do not require all these things all the time: they can tolerate periods of scarcity, some of them having the power of changing into resistant forms in which they can remain for long periods of suspended animation until better conditions set in.

AIR SUPPLY

The volume of air in the soil is governed by the amount of water present. Even after drainage has ceased the pores may retain so much water that only little space is left for air—about 2 per cent on the heavy Gault clay at Cambridge. In old grassland soils rich in organic matter more space is left but this quickly shrinks when the land is broken up for cultivation: in A. J. Low's experiments at Jealotts Hill the space left for air fell to one half after 4 years (Table 1.)

TABLE 1[1]

Percentage of pore space in old grassland and on similar soil after periods of arable cultivation. Division of pore space between air and water when soil is fully moist (field capacity).

	TOTAL PORE SPACE	PART OCCUPIED BY WATER	AIR
Grassland about 100 years old	55·6	35·3	20·2

[1] A. J. Low, *Empire Jl. Expt. Agric.* (1954) 5, 57–74. Figures amended: private communication.

| | TOTAL PORE SPACE | PART OCCUPIED BY | |
		WATER	AIR
Part of above after 4 years cultivation ..	55·3	47·5	7·8
After 4 years cultivation following 25 years			
grass	47·8	38·7	9·2
Continuous arable within living memory ..	41·8	36·8	5·0

After a prolonged drought there is room for much more air. In a normal moist state at Rothamsted air occupied about 7 per cent of the volume of a pasture soil; this increased to 25 per cent after a period of drought: for the unmanured arable soil the values were 11 and 17 respectively.

The composition of the soil air depends on the rate at which it is renewed: the plant roots and soil organisms take up the oxygen and give out carbon dioxide. The quantity of carbon dioxide evolved is considerable: it varies according to the season and the cropping. In autumn it has been 2 to 6 grams per day (1 to 3 litres at normal temperature and pressure) per square metre of soil (1·2 square yards) on bare ground and up to 20 or even 50 grams per day on soils on which plants were growing. But the rate of renewal is more rapid than would at first be expected: in Romell's experiments in Sweden the air in the soil to a depth of 5 inches was completely renewed every hour. The chinks and crannies riddling the soil appear extremely minute to us and it seems incredible that any air could enter. But everything in Nature is relative: the molecules of oxygen are vastly smaller: crevices only $\frac{1}{1000}$ of an inch wide, and far too small for us to see, are as large compared with an oxygen molecule as a valley about 120 miles wide in comparison with a man. For a proper appreciation of the soil one must try to see things as they would appear to its smallest inhabitants if they had the power of sight.

Diffusion is facilitated by passages made by earthworms and other soil animals and by cracks on heavy clay soils: these are of special importance because on such soils the pores are so small that they only empty out their water with great difficulty so leaving little air space. The cracks greatly increase the area of the soil exposed to the air and so facilitate its entry.

In consequence of this rapid rate of diffusion there is on the average little difference in composition between atmospheric and soil air. There are however considerable causes of local variation. Plant roots take in oxygen and give out carbon dioxide just as men and other

animals do, and in their immediate neighbourhood the composition of
the soil air differs considerably from that further away. On grassland
the percentage of carbon dioxide may exceed 3 as against 0·03 in the
atmosphere; the oxygen may fall below 17 against nearly 21. The air
in cultivated soils, however, is much nearer in composition to that of
the atmosphere except when the soil has recently received considerable
dressings of easily decomposable organic matter such as fairly fresh
farmyard manure. The results of a number of samples of soil air taken
systematically at Rothamsted throughout a year at a uniform depth of
6 inches and with precautions to avoid contamination with atmospheric
air are given in Table 2.

TABLE 2

Mean percentages of oxygen and carbon dioxide in the soil air, Rothamsted.

| | USUAL COMPOSITION | | CARBON | |
	CARBON DIOXIDE	OXYGEN	DIOXIDE MAXIMUM	OXYGEN MINIMUM
Arable land, uncropped and unmanured	0·1	20·7	0·4	20·5
Arable land, cropped and unmanured	0·2	20·4	1·4	18·0
Arable land, cropped and dunged	0·4	20·3	3·2	15·7
Old Grassland	1·6	18·4	3·3	16·7
Atmospheric air	0·03	20·96		

Excepting in the layer close to the surface the soil air is almost saturated
with water vapour: this has important effects on the various members
of the soil population. The variations in amount of carbon dioxide are
much greater than those of oxygen and the local variations are almost
certainly greater still. They may be expected to affect the life in the
soil as carbon dioxide is toxic to many organisms; little, however, is
known about this.

There is a marked seasonal effect. The carbon dioxide was at a
minimum in winter but it jumped up in April and May; it fell in
summer then rose to a second peak in August and September; afterwards it fell rapidly again. These results suggest considerable outbursts
of activity of the soil inhabitants in spring and autumn.

In addition to this free atmosphere there is another dissolved in the soil moisture, presumably also in the soil colloids; it contains carbon dioxide and nitrogen only: no oxygen, all having presumably been taken by the soil organisms. This however, is periodically displaced by rain water, which is usually saturated with oxygen.

The soil air at lower depths contains more carbon dioxide and less oxygen than that nearer the surface.

THE WATER IN THE SOIL

When a light shower of rain falls on a dry soil the water soaks downwards for a certain distance and then stops almost as soon as the rain ceases; if you make a clean stroke with a spade you see a sharp line where the penetration ceases and below which the soil remains dry. If no more rain comes little further movement of the water takes place. If however the rain continues long enough and heavily enough the soil becomes completely wet and water drains away into the subsoil. This movement of the water may go on long after the rain ceases as can easily be seen on a drained soil where the pipes open out into a ditch; the flow of water may continue for days even though no more rain falls: indeed from some silty soils a trickle may go on for weeks. When the flow ceases the soil is still wet; a clay may be very wet.

Water that soaks into the depths of the soil does not come back again. There is, it is true, a small rise of water into the soil immediately above the water table, brought about by the same surface forces as filled the pores when the water was descending. It was at one time thought that this capillary attraction would carry the water up to the surface by something analogous to a system of capillary tubes, but this idea has long since had to be abandoned. Experiments at many centres have shewn that the capillary rise is very limited and rarely extends beyond a few feet. This belt is called the capillary fringe; it is of little importance for the supply of water to the plants unless it happens to be so near the surface that the plant roots can get into it.

Water thus exists in two conditions in the soil. Some is sucked up into the pores and held there immoveable with a tenacity that varies with their size and shape and with the quantity of water, while the excess beyond what can be so held is readily moveable provided there is no physical obstacle and it sinks down into the subsoil or finds its way into the streams. The critical pore size is an equivalent diameter of about 30 μ; from larger pores the water can drain away, from smaller ones it cannot. There is no physical distinction between the moveable and the immoveable water: the one is held with a force less than gravity, the other with a greater force. There is, however, a difference that considerably affects plant growth: the immoveable water fills only the smaller pores but leaves some of the larger pores and passages free for air to enter, so satisfying a vital need for plant roots and many soil organisms, while the moveable water may fill all the pore space leaving no room for air, in which case plant roots and organisms may be asphyxiated.

A soil which has just shed all its drainable water, and still retains its maximum quantity of immoveable water, is said to be at field capacity. In this state a suction of about 20 to 40 inches will draw the water out of the larger pores, but as they become emptied a stronger and stronger pull becomes necessary to extract any more. Roots of plants accustomed to humid conditions can exert a pull up to about 10 atmospheres, or 300 ft. of water; beyond this they cannot suck in sufficient water for a continuance of growth or even to keep the stems and leaves erect. Wilting begins. The roots can still, however, imbibe sufficient water to keep the plant just alive although badly wilted, and they can do this even when a suction of 20 to 30 atmospheres is necessary. More they cannot do and the plant dies. The maximum quantity of water that a soil can hold for delivery to the plant is the difference between the amounts present when drainage just ceases, i.e. field capacity, and when wilting just begins; on the Broadbalk unmanured soil at Rotham- sted—a heavy loam—it amounts to 2 inches of water per foot in the top 9 inches, i.e. 22 per cent of the volume, but only about half of this in the subsoil. It is increased by addition of organic matter: on the plot receiving annually 14 tons of farmyard manure since 1843 it amounts to 2·7 inches per foot—30 per cent—not a great difference for the enormous addition made since the experiment began.

There is still moisture left in the soil, however, even when the roots have extracted all they can. Some of this will dry off if the soil is exposed in a thin layer to the air, but a residue remains which can be removed only by heating in a steam oven. At this stage the suction pressure is of the order of 10,000 atmospheres, corresponding to a water column 60 miles high.

These values are not absolute: those for a drying soil differ from those obtained when the soil is being wetted owing to a hysteresis effect, but they show the magnitude of the forces involved.[1]

The amount of water inaccessible to the plant depends on the proportions of clay and organic matter present. Sandy soils keep back very little; loams rather more but not an excessive amount; clay soils more still and fen or peaty soil most of all: in their case the mechanism of retention is not well known. A knowledge of these stages in the relationship of soil moisture content is very important in irrigation practice. There are conditions when it is desirable to flood the soil, but in general it is essential not to add water in excess of the field capacity. Any such excess at best simply drains away and is lost: in various circumstances it can do harm. During the period when the plant should be growing the soil must not be allowed to dry too much. There are, however, no breaks in the curve for the extraction of water by plants or its loss from the soil in any other way, and no "constants" in any strict sense of the word.

Evaporation of water from the soil may cause serious loss, which, however, is generally less than might be expected. R. K. Schofield and H. L. Penman at Rothamsted have shown that the rate of evaporation is nearly the same as from a free surface of water if the conditions are such that the surface of the soil is continuously moist, as is for example the heavy soil at Rothamsted during winter, or the sandy loam at Woburn if a water table is maintained within about 10 inches of the surface. But under warmer conditions at Rothamsted, or with a lower water table at Woburn, the water evaporates from the surface more rapidly than it can be renewed from below, and a dry surface layer forms which even if not very thick, protects the rest of the soil sufficiently to reduce greatly the loss by evaporation. Even after six weeks of the drought of 1921 a fallow soil at Rothamsted had lost only one inch of water: half of this had gone in the first five days when evaporation averaged 0·1 inches daily; later the rate of loss fell to about 0·05 inches a week and when after three months the drought ended the loss had been only 1·3 inches of water. There was no evidence of any appreciable movement of water from the subsoil to the surface as drying proceeded: if any at all moved up it was not more than about a quarter of an inch and that might have been transferred by distil-

[1] Scientific workers often find that the changes in mean free energy of the water are more fruitful for investigation than the suction pressures. The variations are the same for pure water, but differ when the soil water contains salts in solution. For our present purpose the suction phenomena are sufficient.

lation. This slowness of drying arises from the circumstance that it can take place only where the soil water is in contact with soil air as in the larger pore spaces or cracks or at the menisci[1] in the necks of the sacs. In a clay soil the sacs are so small, albeit very numerous, and likewise the necks and menisci, that little or no evaporation from them is possible. Another factor comes into operation in clay soils, however: they shrink on drying, and in doing so reduce the size of the sacs and squeeze out some of the water from them: this gets into the cracks or channels and so evaporates.

The shrinkage of a heavy soil is quite marked: at Rothamsted a gauge based on a foundation 10 ft. deep in the soil has been set up to measure the changes in level due to wetting and drying: differences as great as 2 inches have been recorded. As shrinkage occurs at each crack the change in volume may be considerable. This accordion-like property of clay soils leads to some unexpected results.

The growing plant causes considerable losses of soil moisture. During its growth a plant may transpire some 30 to 50 lb. of water for every lb. of its fresh green weight. In the Rothamsted soil gauges the Woburn soil turfed with grass lost water by transpiration at a rate about three quarters that of evaporation from a free water surface even when the water table was 24 inches below the surface, while bare soil lost very little. A striking comparison was made at Rothamsted after a prolonged drought in 1871. The amount of water was determined in successive 9 inch layers of soil down to a depth of $4\frac{1}{2}$ feet on a bare soil that had been kept free of all vegetation and on adjacent land carrying a barley crop. The results are shown in Fig. 1 and in Table 3: they are expressed in volumes, these being more informative than weights.

TABLE 3

Volumes of air and of water per 100 volumes of soil in the different depths after prolonged drought at Rothamsted, 1871.

	TOP 9 INCHES		2ND 9 INCHES		LOWER DEPTHS	
	FALLOW	CROPPED	FALLOW	CROPPED	FALLOW	CROPPED
Vol. of water ..	25	19	45	30	50	36
Vol. of air ..	15	25	5	14	none	8

falling to zero at about 5 ft.

[1] A meniscus is the curved surface made by a liquid in a narrow tube or neck.

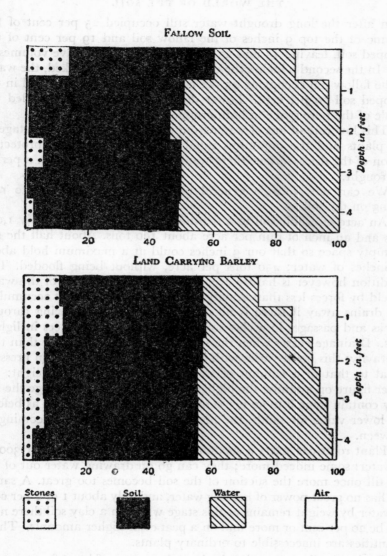

FALLOW SOIL

LAND CARRYING BARLEY

Stones   Soil   Water   Air

FIG. 1

Volumes of soil, water, and air at different depths of a fallow and of a cropped soil at Rothamsted on June 27, 1870, after a prolonged drought. (Calculated by R. K. Schofield from data by Lawes and Gilbert).

Even after the long drought water still occupied 25 per cent of the volume of the top 9 inches of the fallow soil and 19 per cent of the cropped soil, leaving 15 and 25 per cent for the respective volumes of air. In the second 9 inches the differences were also marked, the water in the fallow soil occupying 45 per cent of the total volume and in the cropped soil 30 per cent. Lower down however the water filled the whole of the pore space in both cropped and fallow soils.

This low rate of loss of water from the soil is of great advantage to the plants and the organisms inhabiting it; without this protective action of the dried surface there would be many casualties in a period of drought.

We can now form a general picture of what happens to rain falling on the soil.

An acre of arable soil to a depth of 9 inches weighs roughly 1,000 tons and an inch of rain per acre about 100 tons. About half the soil is empty space so that our 9 inches could at a maximum hold about $4\frac{1}{2}$ inches of water; 450 tons per acre, without being flooded. This condition however is hardly ever reached because much of the water is held by forces less than gravity and it therefore does not accumulate but drains away if there is no obstacle, the drainage being through cracks and passages in a clay soil, and through pore spaces in lighter soils. Drainage ceases when the suction of the soil is greater than that of gravity: this point varies with different soils from a suction pressure equal to that of a column of water 20 to 40 inches in height: the lower figure on a sandy, the higher on a clay soil. At this stage the soil may contain some 8 to 30 per cent of water by volume, and as before, the lower value for sands, the higher for clay, with loams coming in between.

Plant roots can exert a suction of about 10 atmospheres or 300 ft. of water: some indeed more; they can go on drawing water out of the soil till once more the suction of the soil becomes too great. A sandy soil has no great power of holding water, and only about 1 or 2 per cent of water by weight remains at this stage while on a clay soil there may still be 20 per cent or more and on a peat even higher amounts. These quantities are inaccessible to ordinary plants.

On bare soil evaporation is the only source of loss of water once drainage from the surface layer has ceased. It is fairly rapid at the surface but a thin dried layer forming there protects the water in the body of the soil from further loss and subsequent changes are only slow in our climatic conditions. Covering the soil with a mulch of vegetable matter also reduces the loss of water by evaporation. On uncropped land at Rothamsted about 75 per cent of the rain water reaching the soil drains away and 25 per cent evaporates in winter; about 25 per cent drains away and 75 per cent evaporates in summer; while over the whole year about half drains away and the other half evaporates. On cropped land, however, little if anything drains away in summer and about 20 to 25 per cent over the whole year. During their growth plants transpire many times their weight of water; whatever does not come from rain falling during growth has to be taken from the soil and this accounts for the difference in moisture content between cropped and uncropped soils.

The result of these various factors is that the percentage of water in soils varies within limits depending on the percentage of clay and of organic matter. Numerous determinations in three different soils at Woburn not far apart and under similar rainfall conditions gave results recorded in Table 4.

TABLE 4

Range of water content in different soils under similar rainfall conditions at Woburn.

	SANDY SOIL	LOAM	CLAY SOIL
Per cent of clay	5	9·3	43
Range of water content per cent by weight	1 — 14	6 — 16·5	16 — 35
Mean of all determinations ..	9	12	27

THE TEMPERATURE OF THE SOIL

Practically all the heat of the soil comes from the sun: an unknown amount comes from the action of micro-organisms and from radio-activity but so far as is known this is very small. The soil does not receive all the heat radiated from the sun; some is absorbed and scattered in the upper atmosphere but clouds intercept much more: about 75 per cent gets through to the surface when the sky is clear but only about 20 per cent when it is completely cloud covered and 50 per cent when it is half covered. The loss at Rothamsted is shown in Fig. 2 where the upper curve shows the radiation that would have been received had all the days been cloudless, and the lower curves show that which actually reached the surface of the soil.

We are here concerned only with the amount of radiation that reaches the soil: at Rothamsted it is measured daily and the figures expressed in calories are given in Table 5.

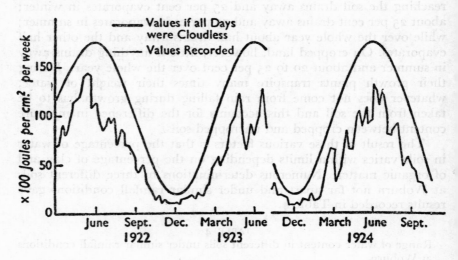

FIG. 2

Solar radiation received at Rothamsted (from *Soil Conditions and Plant Growth*)

TABLE 5

Numbers of calories[1] reaching the surface of the soil at Rothamsted: daily average, and period totals: 10 year period, 1931–1940

	MARCH AND APRIL	MAY TO AUGUST	SEPTEMBER AND OCTOBER	NOVEMBER TO FEBRUARY	WHOLE YEAR
Calories per sq. ft. per day: thousands	216	349	165	59	
Total for period per sq. ft. millions	13	41	10	7	71
Per acre, thousand millions	571	1,774	439	306	3,090

[1] Gram-calories or small calories as used in physics and scientific work generally. Physiologists, nutrition experts and engineers, however, use the kilogram or great calorie, which is 1,000 small calories. Where food production per acre is discussed in this book the results are given in great calories: of these about 3 thousand million are received per acre per annum.

The acre value for the whole year, 3 thousand million great calories, is equivalent to $3\frac{1}{2}$ million units of electricity (kilowatt hours), or $4\frac{1}{2}$ million horse power hours, or to the heat generated by burning some 400 tons of coal.

The amount of radiation received per square foot of soil surface depends on its slope: if it is facing the sun at right angles to the direction of the rays it gets the maximum amount. In Britain a south slope is the warmest position partly because of this effect, partly also because it affords some shelter from the northern winds. Sloping land, however, is not without its dangers. The air at the upper part on cooling rolls down and accumulates wherever the contour of the land retards its further movement. In late spring and early autumn these cold streams of air may be below freezing point and may do considerable damage to plants.

The heat reaching the surface of the soil is not all absorbed: part is reflected back into space. The absorbed portion is not all retained: part is radiated into space; this radiation goes on night and day, while the reception of the heat from the sun is confined to the hours of daylight. The retained portion is largely used up in evaporating water either directly or indirectly by transpiration through the plant; the remainder—only a small fraction—heats the soil and the air, and provides the plant with the energy needed for synthesising its structure materials.

H. L. Penman has drawn up a provisional balance sheet showing how the calories received are on the average expended in summertime in south east England, the soil being in full moist condition.

TABLE 6

Provisional balance sheet: day and night expenditure of energy in summer; soil fully moist, south-east England. Income — 100

Reflection	20
Radiation	34
Evaporation and Transpiration ..	39
Heating air	4
Heating soil	2[1]
Plant growth	1 or less
	100

[1] This is a long term average including an early summer increase, and a late summer fall in temperature; also daily fluctuations. Over the whole year the net gain is *nil*.

The amount reflected by any particular soil depends on its surface. On a sunny day a white chalky soil reflects more than a black one and therefore tends to be cooler; yellow, brown and red soils come in between. A cropped soil reflects more than a bare one, and a bare dry one more than a bare wet one: in some Swedish experiments the bare wet one reflected only about 7 to 10 per cent, the bare dry about double, and the cropped about three times this quantity.

The radiation of heat from the soil back into space presents some very interesting problems. The radiant energy coming from the sun includes a wide range of wave lengths from the very short ultra-violet down to the long heat waves: the very short and the very long ones are absorbed in the atmosphere; what radiation does get through is chiefly of the visible and near infra-red range with a peak wave length of about 0.5 μ. The heat radiated from the soil has a somewhat similar set of wave lengths but the peak is about 10 μ. On a clear night much of this passes out into space, but if the atmosphere is humid the water vapour absorbs certain wave lengths and radiates this heat out again: some of it gets back to the soil. Clouds, which are really masses of minute water droplets, absorb still more of the radiation from the soil, and re-radiate it, also as long waves. A complete cloud cover returns a great part of what the ground has lost: mist and clouds thus act as a blanket protecting the earth from too great a loss of heat.

Some of the radiation absorbed by the soil is used for heating the air, which is done at the soil surface. By far the greatest part, however, is used for evaporation and transpiration of water: this is more fully dealt with later on. (p. 180). The rest of the radiation is chiefly used in warming the soil, and the extent to which it can do this depends largely on the amount of water present. A very wet soil requires more than twice as much heat as a dry soil to raise its temperature through a given range because the specific heat of water is three or four times that of dry soil. Moisture facilitates the flow of the surface heat into the soil, its conductivity being greater than that of the air it has displaced, but in evaporating it takes up a large amount of heat.

The heat taken up by the soil raises the temperature of the surface layer during the day but during the night the temperature falls. If the night is clear and free from mist or cloud and the soil is dry the daily variation is considerable in hot weather: in summer at Rothamsted it may be about 35°C. But if the soil is wet the daily variation is down to about 20°C because of the protection afforded by the water vapour. The heat retained by the soil travels downwards but so slowly that even at 2 inches depth the temperature during the day never rises as high as at the surface nor does it fall so low during the night: the daily variation is heavily damped down; at 8 inches depth the range is very small. Between summer and winter there is a greater temperature variation but it too is heavily damped down in the body of the soil compared with the variation at the surface. During summer the heat moves downwards, but during winter upwards, and this continues into the spring when heat is also beginning to move downward: there is a similar double movement in autumn. The net result is that for plant roots and organisms at 8 or 12 inches depth there is nothing to show any difference between day and night, or any great difference between summer and winter. In this respect Rothamsted is probably typical of much of England.

On the sandy soil at Woburn the temperature wave moves more quickly than at Rothamsted; the soil cools more rapidly in winter and warms more rapidly in spring, but in autumn the mean temperatures are the same at both stations.

A vegetation cover whether living or dead modifies the temperature relationships by damping down the fluctuations especially at the surface. It protects the soil considerably against frost in winter. Fig. 3 shows that during a cold spell in the winter of 1939-40 the January frost penetrated sharply to a depth of four inches in a bare soil but not in a turfed soil; the temperature at four inches depth under turf was no lower than at eight inches depth in the bare soil. During the growing season the plant leaves absorb much of the radiation so that it does not reach the soil: the leaves and the surrounding air are therefore warmer than the soil by day: at night however the leaves radiate out some of their heat while continuing to blanket the radiation from the soil: the leaves and surrounding air are therefore cooler than the soil, this is specially marked on dew nights when the soil surface may be 4°C warmer than the air in the grass cover at a height of 1 cm. (0·4 inches) above the surface and this leads to a distillation of water from the soil on to the grass.[1] Over bare soil the conditions are reversed: the air temperature is lower than that of the soil surface during the day but higher during the night. The climate inside the crop thus differs somewhat from that outside; it is called the micro-climate.

[1] J. L. Monteith, Rothamsted Annual Report 1954, p. 37.

The differences have important effects on the incidence of plant and animal diseases.

SOIL REACTION

Plants and other organisms inhabiting the soil are subject to a condition which vitally affects their growth and even their survival, but of which human beings and animals on the surface of the soil have no experience and can only dimly imagine. The soil may be either

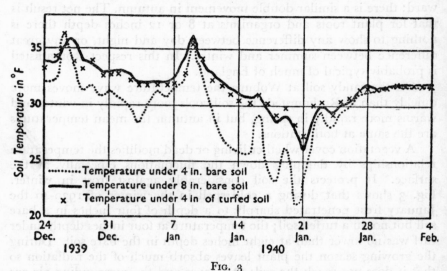

FIG. 3

The effect of a turf covering on the sub-surface temperature of the Rothamsted soil during a hard frost (from *Soil Conditions and Plant Growth*).

acid, neutral or alkaline: there is an unbroken range of possibilities like the sounds of a violin, and for each plant and organism there is an optimum region on the scale outside of which it fares badly and if too far away it dies. The scale is measured by the hydrogen ion concentration, but as the values in the ordinary units are inconveniently large the logarithm of the reciprocal, i.e. the negative logarithm, is used instead and the numbers are called pH. It is not necessary to go into the rather complicated basic theory: the scale can be accepted as having been well tested by experience.

Plate I. Growing complexity of soil science: *above*, the first Rothamsted Laboratory (1843); *below*, one of the soil laboratories (1956)

Plate II. First stage in soil formation; wind, water and frost break up the rocks. Elan Valley, Radnor.

The value 7 stands for neutrality: higher values indicate alkalinity, and lower values acidity; as the numbers are logarithms a difference of 1 is equivalent to multiplying by 10, and a difference of 3, multiplying by a thousand. In Britain and other countries of temperate climates and moderate rainfall it is rare to find alkaline soils of pH higher than 8: the theoretical maximum for soils in which calcium is the dominant exchangeable cation—a usual condition in good English soil—is 8·5, but in dry climates sodium carbonate may be present in the soil and then the value may rise higher: the highest recorded is 10·0 in a soil in Washington State. British soils more usually tend to be on the acid side, about 6·4 to 6·9: a value of about 4·5 is as low as one usually finds on mineral soils but lower values due to organic acids are found on peats; an uncultivated part of Carrington Moss had a value of 3·0 only.

The acidity is due partly to some of the organic matter and partly to some of the clay minerals. Some of the organic substances contain carboxyl and phenolic groups from which hydrogen ions can dissociate: they are therefore true acids. It was formerly thought that the clay acidity was likewise due to hydrogen ions which had taken the place of calcium and other replaceable ions in the clay particle and which dissociated like those in any other weak acid. Later work has thrown doubt on this view: a hydrogen clay may be formed, but it is very unstable; the lattice disintegrates liberating aluminium ions which then act as replaceable bases and neutralise much of the negative charge on the lattice. Acid clays are thus mixed aluminium-hydrogen clays and not simply hydrogen clays: apparently they contain no other exchangeable base, and in particular no calcium. How they come to possess acid properties is too complex a subject for discussion here.[1]

The acidity of a soil is not a fixed quantity: it depends to some extent on the conditions. It tends to increase somewhat as the soil dries. Also the reaction of the plant to the soil acidity varies with the conditions; it is less sensitive, and can tolerate more acidity, i.e. a lower position in the pH scale, in cool wet conditions than in those warmer and drier; lower in Wales for instance than in the Eastern Counties. The pH does not, however, fluctuate rapidly: the humus and the clay colloids act as buffers damping down the effects of factors that might be expected to cause change.

[1] It is fully dealt with in *Soil Conditions and Plant Growth*, 9th Edition revised by E. Walter Russell, Longmans 1961.

WOS – E

CHAPTER 4
THE SOIL INHABITANTS:
I. THE INVISIBLES

THE PICTURE of the soil that emerges from the preceding pages is very different from what one might at first imagine. It is far from being the dense solid mass that it appears to be; only about half its volume is solid matter, the rest is space, empty except for air and water. But the space is so broken up into pores, minute passages and fissures invisible to our eyes that we fail to appreciate its true extent. The smaller of these spaces are filled with water, the larger with air; this contains nearly as much oxygen as atmospheric air because diffusion readily goes on; it contains, however, more carbon dioxide and it is saturated with moisture. The water contains dissolved mineral and organic matter much of which has nutritive value; it is renewed by the rain which comes in saturated with oxygen and permeates the whole soil mass so far as it can go. The solid part of the soil is a mixture of mineral and organic matter, both containing nutrient elements and the organic matter contains also sources of energy; it is in the form of grains coated with a jelly composed of mineral and organic substances.

The temperature in the body of the soil is more equable than in the air, being neither so cold in winter nor so hot in summer; except just at the surface the soil is well lagged against changes. Occasional harmful substances are produced by chemical changes in the soil—by reduction when oxygen is excluded, and by washing out of lime by rain, resulting in acidity—but these changes are exceptional and in any case the soil is well buffered against changes in acidity. Air, water, food, warmth, energy supply, absence of seriously harmful factors, all the conditions for life are there: only one factor that we enjoy is missing —sunlight. In accordance with the general rule that life appears wherever it is possible a wonderful population of living organisms has developed in the soil, adapted to the conditions there and not at all hampered by the absence of sunlight; indeed so little are they suited to it that for many it is fatal.

Recognition of this population came late and even now but little is known about it because of the great difficulties of investigation: you cannot as yet put a fragment of soil under a high powered microscope and explore the fissures and the crannies to see what is there. Special staining techniques have been developed, however, by which something can be learned and one need not despair; some day it may be possible to make a film showing a day in the life of the soil population: until then most of our knowledge is by inference.

Agricultural science has often benefited by military necessities, and the beginnings of the idea of a soil population go back to the 18th Century when the French, needing an assured supply of saltpetre for making gunpowder, arranged for a systematic study of the nitre beds in which it was generated. These consisted of heaps of soil, excrements, and vegetable matter with some wood ashes kept under cover so that the heat generated by the decomposition should not be lost. Stable drainings, urine and like liquids were added from time to time so as to keep the heaps moist. After a time the heaps were well lixiviated with clean water to wash out the saltpetre that had been formed; the liquid was evaporated and the saltpetre crystallised out. Its alternative name was "nitre", and the process became known as "nitrification". Chemists soon found that the nitre—nitrate of potash—had been produced by oxidation of ammonia resulting from the decomposition of the organic substances in the heaps, but they could not understand how the oxidation was achieved; by no methods could they effect it in the laboratory. It remained for more than a century one of the mysteries of science.

The solution of the mystery came from quite an unexpected quarter, as has often happened. Two French chemists, Schloesing and Müntz, were asked by the Paris Municipality in 1877 to study the purification of sewage. They found that when the sewage was allowed to trickle continuously through a column of chalk it passed through unaltered for some days but after that a remarkable change set in: the sewage began to come through pure, and it now contained nitrate. Purification, they saw, involved nitrification. But why was there a delay at the outset?

It was the time when Louis Pasteur was doing his magnificent work in Paris on the activity of micro-organisms, and whenever some inexplicable phenomenon was observed the question was asked: might micro-organisms be the cause? If so, the delay was easily explained: the organisms would be brought in by the sewage but time would be required for the building up of a sufficiently large population in the tube to effect the changes. If, on the other hand, the process were chemical or physical there should have been no delay. To test their hypothesis they blew chloroform vapour through the tube and found that this stopped all nitrification; then they blew out the chloroform vapour and found that the process started again. This was convincing proof that nitrification was brought about by micro-organisms living in the soil.

Schloesing went a stage further. If this explanation were correct the organisms must be taking in oxygen and giving out carbon dioxide. He showed that this was happening; oxygen steadily disappeared from a bottle containing soil, hermetically sealed and fitted with a tube dipping under mercury, but there was no disappearance when the soil had been sterilised with chloroform.

Agricultural chemists by this time knew that nitrate was produced in the soil from farmyard and other organic manures and was indeed one of the chief reasons for their value; however the production was effected it was a vital factor in the growth of crops. And now it appeared that this vitally important process on which human beings depended for their very existence was brought about by mysterious organisms living in the soil. They could not at first be identified, but some were picked out; they were extremely small, visible only under a good microscope and then only as dots or short rods. Pasteur and other investigators had shown that similar organisms produced diseases, others caused the souring of milk, of beer and wine and a variety of important changes. It is impossible for us in this generation who have almost lost the sense of wonder to realise the astonishment mixed with awe with which these discoveries were received. Percy Frankland and his wife wrote a delightful little book: "Our Secret Friend and Foes", which set out in beautifully simple language what was known about this mysterious new world of invisible living things around us, some absolutely indispensible to us, others useful, others again indifferent, or even malignant. It is a fascinating volume which even now can be read with pleasure and it had a wide circulation.

This is not the place to trace the later history of the subject. It is a story of deep human interest; of men and women who, like Browning's grammarian, devoted their whole lives to the study of some apparently remote problem, yet failed to achieve success: sometimes suffering the bitter disappointment that a new-comer, by varying the technique, solved the problem in a few months—sometimes, too, the tragedy has been heightened by the discovery years afterwards that their observations, recorded, but little heeded at the time, contained information of great value for the solution of some other problem.

Of all the many kinds of organisms found in the soil the most remarkable are the bacteria. They are the furthest away from all our experience as regards their size, their mode of life, their method of reproduction and their capacity to change. Ordinary English measures of size are quite inapplicable: the unit of measurement is the micron, one-thousandth of a millimetre and denoted by the Greek letter mu— μ—it is about $\frac{1}{25000}$ of an inch: the organisms range from 1 μ in length and 0·15 μ diameter upwards. Many attempts have been made to form a mental picture of what these sizes mean. One of the best was S. G. Paine's: he calculated that a quarter of a million of the organisms could sit comfortably on an area the size of the full stop at the end of this sentence. This seems inconceivably small to us, but everything in Nature is relative; it is large compared with the phages described later, and vastly larger than the molecules of oxygen that rush at high speed in and out of the soil pores, while they in turn would appear colossal to one of the electrons of the physicist. Astronomers tell us that the Universe is continuously expanding and that there is no limit to the infinitely great; it appears that there is no limit when one turns to things diminishing in size, to the infinitely little.

The highest power microscope shows that the organisms consist of a single cell surrounded by a cell wall; some are round, some rod shaped, others spiral, some have thin tails called flagella which enable them to wriggle a way through moist soil. Their shape is not constant, however, but can be varied in the laboratory by altering the conditions of culture: a dot may change to a rod, a small form to a large one, or conversely. The majority of bacteria in the soil, however, are short rods about 0·5 μ diameter and 1 to 3 μ long.

The electron microscope has a much higher magnification and shows more detail of the structure of the cell walls and of the flagella; it reveals also something resembling a nucleus which, however, is not very well defined and may be variable. (Plate IV, p. 67).

The old idea was that reproduction was very simple, the cell merely splitting in two with nothing in the nature of a sexual process. Later genetical investigations have shown, however, that hybridisation, recombination, and "crossing over" all take place, implying some mechanisation like that in the chromosomes of more complex organisms. It is of course the old story: the first observations give the impression of simplicity; later studies with better appliances reveal greater complexity, and as the appliances and the observations improve so the complexity increases and the wonder and the mystery deepen. However it is brought about the rate of multiplication can be extraordinarily rapid; a newly formed cell may split into two in some 20 or 30 minutes. Some hair-raising calculations have been made by ingenious minded persons showing what might happen if conditions remained favourable. A well known German bacteriologist, F. Löhnis, estimated that a single cell could in the course of 25 hours produce a colony 1,000 cubic meters (1,308 cubic yards) in size which would fill 100 waggons of a goods train. Happily this rate of proliferation is unattainable: apart from the difficulty of supplying enough food sufficiently quickly there is the greater difficulty of removing the toxic products of excretion.

In some species a spore can develop inside the cell; it swells, and the cell gradually disappears. The spore is very resistant to heat, cold or dryness; it can remain for years without apparent change, and then, when conditions are favourable, it can germinate and produce a new cell. The spore formers can thus survive long periods of adverse conditions where species unable to form spores would perish.

The method of feeding is, for many bacteria, in principle not unlike ours. For the aerobic saprophytic organisms, i.e. those requiring oxygen and organic matter, the foods are fats, carbohydrates, and proteins as for us. Being surrounded by a cell wall the bacteria are unable to assimilate solid matter: instead they secrete solvent agents called enzymes which break down the food material into simpler soluble substances that then pass through the wall into the cell—a process similar to that taking place in the human stomach. An enzyme is a specialised agent: each is able to tackle a particular type of compound and no other; the enzyme that breaks down starch will not touch cellulose or protein, and conversely the protein-splitting enzymes will not touch starch: the relationship is that of a key to a lock. Besides the breaking down there is also a building up process with formation of complex poly-saccharides containing uronic units, some also containing amino-sugars.

Perhaps the most astonishing characteristic of the soil bacteria is the wide range of organic substances they can break down. Carbon ring compounds like naphthalene and toluene, decomposable by chemists only at high temperatures and with powerful agents, are readily broken down by soil bacteria at ordinary temperatures and with no adventitious aids, and used by them as sources of food and energy. How they do it is a baffling mystery: never in their natural life can they have come across some of these compounds. It was this that made the preparation of a good soil insecticide so difficult: various agents poisonous to insects and suitable in other ways for practical use were promptly decomposed by the soil bacteria. It was found, however, that the introduction of chlorine atoms into the molecule protected the substance against bacterial decomposition and that method is now adopted. Protection is not complete however; both o-chloro- and o-bromo-phenol are decomposed, but not p-chlorophenol, though p-chlorophenoxy-acetic acid is to some extent, and this action can be intensified by careful nursing of the organisms: repeatedly adding small but increasing doses of the compound. It is not known whether this is the result of building up a large population of a specific organism or group adapted to these particular compounds, or whether a change is induced in some of the organisms giving them this new power of attacking and decomposing a substance formerly poisonous to them.

Some of the bacteria of fertile soils have rather complex requirements, needing growth-promoting substances; some of these compounds have been found in water extracts of fertile soils, though they have not been identified. On the other hand the bacteria in infertile soils have simpler requirements or greater powers of making the growth-promoting substances they require; at any rate these substances have not been found in the water extracts of the soil. The distribution of the species of bacteria does not seem to depend on the fertility of the soil, however; the same species are dominant on all the Broadbalk wheat plots at Rothamsted so far as they have been examined: on the very exhausted plot cropped without any manure or fertilizer since 1839, and on the very rich plot that has received annual dressings of 14 tons per acre of farmyard manure since 1843. But a different set of species is dominant on Barnfield, which is cropped with mangolds; whether this is due to the difference in crop and therefore of root action (p. 88) or to the difference in soil management necessitated by the differing crop requirements, is not known.

Some of the bacteria have the peculiarity that they do not use the free oxygen of the atmosphere, as most other living things do, but instead take it from some compound, usually a nitrate or a sulphate. They are called anaerobic because they can live without air. Those bacteria that, like ourselves, require air, are called aerobic.

In studying any scientific problem it is always necessary to devise some means of measuring the quantities concerned: in this case the numbers of bacteria in a given weight of soil. The first method adopted was to shake up a weighed quantity of soil with water so as to dislodge the bacteria, then after appropriate dilution to pour a measured quantity of the liquid on to a plate of gelatin or agar-agar containing nutrients. The plate was then covered, put in an incubator, and left for a few days. Multiplication was so rapid that each organism had now become a colony visible to the naked eye. These were counted, and knowing what fraction of the original soil had been put on to the plate it was possible to estimate the minimum numbers of bacteria in the soil—minimum, because the plate may not have suited all the different kinds, and those that failed to multiply would not be seen, also because some of the bacteria might be clustered in the soil, and the colony instead of originating from one organism might have arisen from a cluster numbering thousands.

Later improved microscopes and staining techniques enabled direct counts to be made. These gave far higher numbers. The early plate counts were frequently 3 or 4 million per gram—100 million or more per ounce,[1] which 50 years ago seemed tremendously high. As the technique improved the counts rose: 25 or 30 million per gram became common. But direct counts have put these figures completely in the shade: 1,000 million to 4,000 million per gram are now frequently recorded. A saltspoonful of a dry garden soil weighs about $1\frac{1}{2}$ grams; on present knowledge it contains roughly twice as many bacteria as there are men, women and children in the whole world. The figures include not only bacteria but some moribund and recently dead, as well as spores of actinomycetes, a group described later. The two methods do not tally: for soils from differently manured plots at Rothamsted the microscope counts do not run parallel with the plate counts.

[1] There are 28.35 grams in 1 oz.

Nor do they fluctuate in the same way with variations in temperature and moisture supply. The Rothamsted workers suggest that the plate method records an element in the soil population that reacts to external factors differently from the entire population. The plate method is, however, very useful for the isolation of bacteria: the colonies can be picked off the plate and transferred to nutrient solutions for further study.

On the Broadbalk wheat field the numbers as shown by the microscope counts fluctuate with the moisture content and temperature of the soil, falling when the moisture falls and rising when it rises, but behaving in the opposite way towards temperature changes. The numbers are higher on the plot receiving farmyard manure than on those receiving fertilizers or nothing, but they are not consistently higher on the fertilized plots than on the unmanured—differing in this respect from the fungi.

The most remarkable feature of the numbers is that they are continuously varying. Fig. 4 shows the changes in number in samples taken from a fallowed garden plot every two hours day and night from 7 a.m. on the morning of October 24th to 5 a.m. on the morning of October 26th, 1934. The changes showed no relation to the variations in temperature and moisture content; indeed they occurred in soils kept in the laboratory under absolutely constant conditions. They are probably related to variations in the numbers and activities of other micro-organisms, especially the predators to be described later; they show however that the soil population must be in a state of considerable turmoil.

These aerobic saprophytic bacteria play a vital part in the formation of humus and the production of plant food in the soil, though it is not yet possible to assign specific processes to any particular groups of them: indeed it is improbable that their actions are specific, they are more likely to be capable of effecting a variety of changes according to the different conditions.

SPECIALISED BACTERIA OF RESTRICTED ACTIVITIES

Soil also contains some extremely interesting groups of organisms which are much more specific in their requirements and their effects, some indeed can bring about one or two changes only, but these include changes of the utmost importance for the welfare of animals

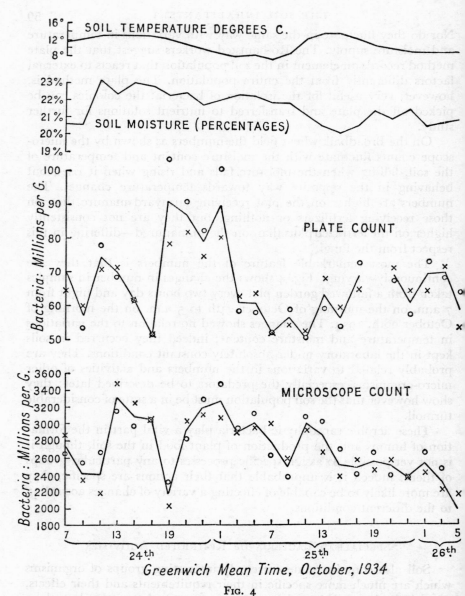

Fig. 4

Counts of bacteria from two hourly duplicate samples of fallow garden soil, Rothamsted.
(H. G. Thornton and G. B. Taylor: from *Soil Conditions and Plant Growth*).

and human beings. Unlike the organisms so far described many of these specialised bacteria do not depend on constant supplies of free oxygen and of organic matter as sources of food and energy, nor do they all need combined nitrogen, i.e. protein, peptone and the like. Some of them can obtain their oxygen by the reduction of an inorganic compound such as a nitrate, a sulphate, a nitrite, or a higher oxide of iron or manganese. Some obtain their carbon from carbon dioxide and derive the energy necessary for the process from the oxidation of a simple inorganic substance like ammonia or even sulphur. Others, most remarkable of all, have no need for protein or indeed any form of combined nitrogen, but can utilise the free nitrogen of the atmosphere. Between them these various groups bring about a supremely important cycle of nitrogen changes.

The Nitrogen Cycle in the Soil: I. Fixation of Atmospheric Nitrogen

Agricultural chemists had long been puzzled by the presence of nitrogen compounds in the soil. How had they got there? None are present in the original rocks, and although nitrogen itself is abundant in the atmosphere it is very inert and can only be made to combine with other substances at temperatures far exceeding those occurring in Nature, such as the electric arc or, at lower but still high levels, by the use of wholly artificial catalysts. Some small amount of combination takes place in the path of a stroke of lightning, but the resulting compound is so easily removed from the soil that it could not possibly persist. Not only is the quantity of fixed nitrogen in the soil considerable—some $1\frac{1}{2}$ to 3 tons per acre in the top 9 inches, and further quantities below that—but as already shown it is subject to continuous loss by the washing out of nitrates by rain, and by other causes. In spite of these losses the quantity of nitrogen in a soil that has reached a steady state remains unchanged. Obviously there must be recuperative processes or the nitrogen in uncultivated soils would have gone long ago.

As has often happened in the history of science the first step in solving the problem was taken in France, much painstaking and systematic development went on in Germany, and there was further clarification of the subject in England. In 1885 Marcellin Berthelot, one of the most versatile and prolific chemists of his day, and working at a time when French scientists were steeped in bacterial lore, exposed sterilised and unsterilised sands and clays poor in nitrogen (0·01 per cent or less) to air in large closed flasks for five months and found distinct gains in nitrogen in the unsterilised, but not in the sterilised soils. Fixation of nitrogen, he concluded, was therefore due to bacteria. Looking back it is very difficult to see how this result can have been obtained because the essential condition for fixation, an adequate supply of organic matter, does not seem to have been provided; one feels that his assistant must have made some mistake. But it would be churlish to be too critical because the investigation was extremely fruitful: it provided the key to the clover problem just when two very able German investigators, Hellriegel and Wilfarth, were working at it, and it stimulated a brilliant young Russian investigator, Serge Winogradsky, to search for the actual organism.

Few scientists of his day showed greater ingenuity than Winogradsky in devising methods at once simple, direct, and effective for the solution of his problems. He realised that the most hopeful way of picking out the organisms from the mixed crowd in the soil was to grow them in a medium as favourable as possible for nitrogen fixers and as unfavourable as possible for the others: he therefore made a medium containing all the mineral nutrients but no nitrogen compounds; no organism dependent on these could grow. Equally important, he realised that nitrogen fixation was an endothermic process and the organisms must therefore be provided with a source of energy: he therefore added sugar to the medium and finally a little soil. Decomposition speedily began; the sugar was decomposed with evolution of carbon dioxide and production of some simple fatty acids, and nitrogen was fixed. Three species of bacteria developed, of which one was a Clostridium. They stuck together obstinately and all efforts to separate them completely failed. Some of the bacteria were dislodged and tested: they could not fix nitrogen. The active agent appeared therefore to be the Clostridium but somehow it had to be freed from its companions.

Here Winogradsky showed his great ingenuity. Up to this stage air had always been present: he now excluded it completely and the bacteria died leaving the Clostridium free. But now the Clostridium would not fix nitrogen when put into the original medium with access to air as before. Nothing daunted, he cut out the air, letting in nitrogen only: fixation went on actively. This showed that the Clostridium belonged to the anaerobic group of organisms that have no need of free oxygen: some indeed cannot tolerate it. Why had fixation taken place in the first instance when oxygen was present? The explanation was that the bacteria so firmly associated with the Clostridium had taken up the oxygen, serving as a kind of protective blanket and allowing the fixation of nitrogen to go on in an anaerobic microclimate.

Before long another nitrogen fixer was discovered: this time by a Dutch investigator, Beijerinck. It is aerobic and more efficient than Clostridium; it can fix 1 part of nitrogen for 20 parts of carbon transformed:[1] it is called Azotobacter. Winogradsky had missed it for a very interesting reason. In order to simplify his problem of reducing the soil population to manageable proportions he had heated the soil to 75°C to kill organisms that did not form spores—and in so doing killed the Azotobacter, one of the very things he was looking for. No scientific technique is ever perfect.

Both Clostridium and Azotobacter are widespread throughout the world, especially Clostridium which is more tolerant of soil acidity than Azotobacter. The chemistry of the process is as yet unknown but the isotope of nitrogen (N^{15}) will help towards solving the problem. One would expect ammonia to be the first product, but if so it must instantly combine with something else for it cannot be detected. The first detectable product appears to be glutamic acid which however could be formed from ammonia by interaction with α-keto-glutaric acid. Aspartic acid is another early product.

No chemist can yet imitate this production of cell protein from free nitrogen at ordinary temperatures with no appliances; no nitrogen fixing enzymes are known. For some reason, equally unknown, a minute quantity of a molybdenum salt is necessary for the fixation: this has only been discovered in recent years and much of the older work is vitiated by the circumstance that the experimenters did not—indeed could not possibly—know of this essential condition. The scientific investigator is working in the dark: as Browning put it, he explores with a taper, not with a torch, and inevitably misses a great deal.

[1] For Clostridium the value is probably only about a quarter of this. But these results refer only to conditions of laboratory cultures: what happens in the soil is as yet unknown. Clostridium is far more numerous than Azotobacter on the Broadbalk soil (Jane Meiklejohn).

We still do not know whether either of these organisms actually fixes much nitrogen in natural conditions and until recently it was impossible to discover, because losses of nitrogen are continually taking place as will be shown later, and only the differences between gains and losses have hitherto been measured. There is some evidence that Clostridium may play an important part in acid forest soils. Fixation does not occur, however, if the organisms have a supply of nitrate. E. W. Russell has suggested that the gains could be estimated if the experiment were done with air enriched with isotopic nitrogen, N^{15}, the changes of which can be accurately followed.

A third group of nitrogen fixers is much more important and there is no uncertainty about the extent of their activity. It has been known since very early times that plants like vetches, lucerne and clover enrich the soil during their growth: this was recorded by Theophrastus in 300 B.C. When in the 19th Century chemists turned their attention seriously to soil problems they soon showed that the enrichment was due to a considerable addition of nitrogen. Lawes and Gilbert proved in 1861 that the growing clover plant had by itself no power of fixing nitrogen: not till Berthellot's experiments in 1885 had given Hellriegel and Wilfarth the key to the problem was it discovered that the fixation is brought about by micro-organisms in association with the growing leguminous plant. If one asks why Lawes and Gilbert missed this discovery the answer is first that agricultural scientists were not thinking about bacteria in 1861, and more important, that the very thoroughness of their technique prevented any possibility of their making this discovery. For in order to avoid any contamination or adventitious complication they had heated the soil on which the clover was to grow, and so carefully cleaned their apparatus and the air and water supplied to the plant that they were in fact working in sterile conditions and maintained them during the whole of the long experimental period. It was a triumph of good manipulation—but it missed the great discovery: another of the pitfalls for the scientist.

Once the clue was furnished investigations of the process of fixation were made in many countries. At Rothamsted H. G. Thornton and his colleagues have systematically studied the whole process in the case of lucerne and revealed a fascinating but complex example of co-operation between a plant and a micro-organism.

The organisms occur in soil as minute dots or cocci which develop flagella and become motile; they then change to rods which after passing through several stages become cocci again. Some of the motile forms are among the smallest organisms known, being only 0.9μ long and 0.18μ wide. (Plate IV, p. 67). The association with the plant begins early in its life, soon after its first true leaves are opened. The plant then excretes from its roots some substance that stimulates the organisms to swim through the soil to the root and multiply there. The speed of travel in favourable conditions is about one inch in 24 hours: normally it may be much less. A colony forms near the tip of a root hair; it excretes a substance—probably indolylacetic acid—which causes the root hair to curl, and at the bent tip the bacteria make a way through the cell walls into the root hair; thence they pass into the cells of the root where they multiply rapidly, so forming the nodule. Vascular strands from the root grow along the sides of the nodule; through these food and energy substances are passed from the plant to the bacteria.

With the source of energy thus provided the bacteria proceed to fix nitrogen, building it up into soluble compounds some of which may be excreted by the roots—Virtanen showed that aspartic acid was so secreted—and some pass along the strands into the plant. The process is thus mutually beneficial: the plant supplies the organisms with carbohydrate food and energy; while the organisms supply the plant with nitrogenous food; it is a perfect symbiosis (see page 75). It breaks down, however, if either partner fails to function and this may happen through a variety of causes. Sugar cannot be produced if the plant is kept in darkness. It cannot get to the nodule unless the connecting strands develop; in the case of broad beans they cannot do this without a minute quantity of boron: the reason is quite unknown. (Plate IX, p. 162). The organisms cannot fix nitrogen unless they are well supplied with calcium and phosphate and have their tot of molybdenum: again the reason is unknown. They will not fix nitrogen if this is already available in the form of ammonia or nitrate. What is even more mystifying, active nodules contain haemoglobin, that characteristic component of the blood of man and higher animals: it is always present in nodules when the partnership is functioning properly though it has never been found associated with other plants or soil micro-organisms, nor does it occur in leguminous plants without nodules, or in nodule bacteria outside the nodule.

If any of these conditions is absent the partnership breaks down, the organisms become pure parasites, and attack the root cells to provide themselves with food and energy without, however, contributing anything to the plant. The chances of a successful partnership look slender, but obviously they are not, as shown by the ease with which clover springs up unwanted in lawns.

A further peculiarity in this remarkable story is that the bacteria concerned are very selective in regard to their plants. The organism that co-operates with clover will not do so with lucerne, lupins, or soy beans; each of these has its own special kind, and they are not normally interchangeable; the cowpea organism in America is, however, more catholic in its combinations. When a new species of leguminous plant is introduced into a farm or garden it is necessary to bring in also the appropriate organism if it is not already there, or alternatively to treat the plant as non-leguminous, giving it a nitrogenous fertilizer or manure. In many parts of Great Britain lucerne—which is not a native plant—will not grow satisfactorily unless its particular nodule organisms are added to the soil: this is so well recognised that some merchants now inoculate the seed by mixing a culture of the organisms with it before sending it out.

A properly selected strain must, however, be used. Different strains of the organisms even of the same group vary in their effectiveness; some can get into the roots but do little when they get there, usually forming only small nodules, and fixing little or no nitrogen. Ineffectiveness is due to incompatability between host plant and organisms: a given strain of bacterium can be effective in some plants but not in others. Red clover has been extensively studied; some strains of its nodule organism are ineffective in all plants. But the plants also differ as hosts; some can produce only ineffective nodules even though the invading bacteria are normally effective. P. S. Nutman has shown that genetically controlled factors in both plants and bacteria seem to determine (1) whether infection can occur at all, (2) the number of nodules formed, and (3) whether these are effective or not.

Plants producing many lateral roots also produce many nodules and *vice versa*, suggesting that infection occurs at certain sites only on

---

Plate III. a. Clay particles, Rothamsted soil, x 40,000. (*Electron micrograph:* H. L. Nixon)
b. Coating on soil particles, Rothamsted. (*Photo micrograph: V. Stansfield*)

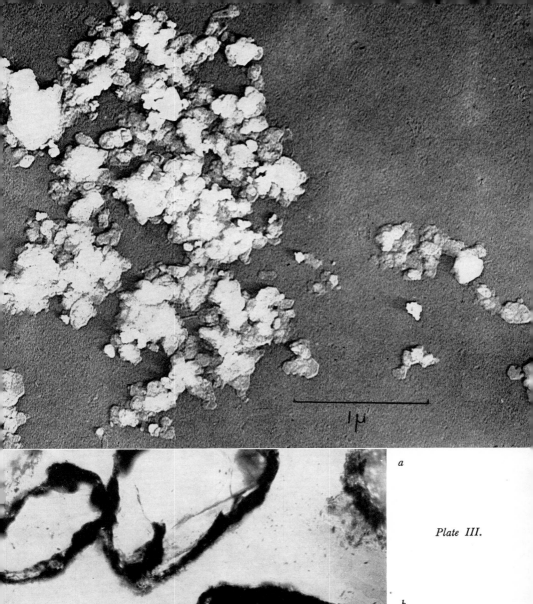

a

Plate III.

b

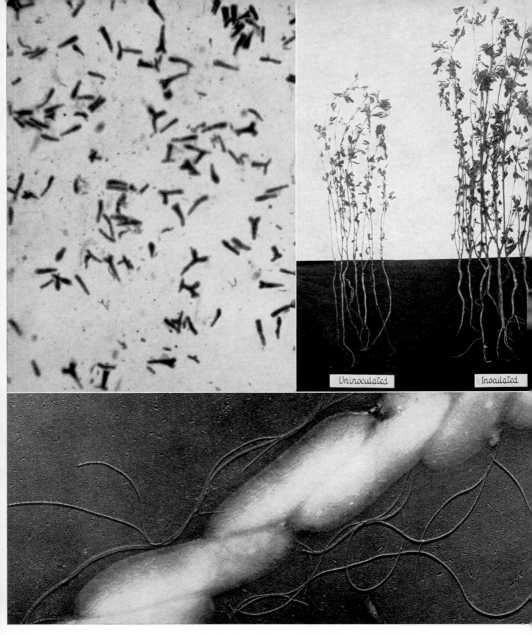

Plate IVa and b. Bacteroids of Rhizobium of Red Clover (*Rhizobium trifollii*) : a (above left), × 1000 (*V. Stansfield*); b (below), × 15,000 (*Electron micrograph: H. L. Nixon*); c (right), Lucerne plants with and without their proper bacteria.

the root. Also the presence of a large nodule reduces the number of subsequent infections, probably the result of some inhibiting substance which it secretes. This interference is not confined to the plant acting as host; secretions from its roots can reduce infection in neighbouring plants.

Like other bacteria the nodule organisms are not rigidly fixed in character and the offspring are not necessarily like the parent. In laboratory cultures an effective strain may dissociate into less effective mutants each of which may continue to produce its like, or it may produce an effective throwback; there is however no record of a naturally occurring ineffective strain producing a completely effective mutant. These ineffective strains are widespread in the hill districts of the west and north of Britain, which no doubt accounts for some of the difficulties of establishing clover there. Attempts to replace them by more effective strains have met with several difficulties: apart from competition by native forms there is a considerable tendency for effective strains introduced into some soils to degenerate and to produce ineffective mutants; any strain developed for purposes of inoculation must combine within itself the three essential properties of effectiveness, genetical stability, and vigour in meeting competition.[1]

The amount of nitrogen fixed by a leguminous crop making good healthy growth is considerable: values of the order of 250 to 350 lb. per acre or more, equivalent to a dressing of 10 to 15 cwt. of sulphate of ammonia, are not uncommon. Some farmers rely entirely on lucerne or the clovers to supply the nitrogen needed for most of their arable crops.

All human, animal, and plant life is completely dependent on nitrogen fixation, and it is a sombre thought that the ability to effect this is possessed by so few and such lowly types of organisms. Nor are they entirely safe in the soil: as will be shown later they are liable to interference or even destruction by some of the other soil inhabitants: actinomycetes, fungi, bacteria and that curious entity phage, though they are not readily taken by the soil amoebae, in which respect they differ from Azotobacter. At present the balance of conditions in the soil is normally in their favour, but if it should shift in favour of their

[1] A full account of these organisms and their relation to the host plant is given by H. G. Thornton in *Science Progress*, April 1954, No. 166, pp. 185-204. P. S. Nutman's genetical studies are in *Heredity*, 1949, 3 and 1954, 8, and in later journals.

antagonists the consequences would be catastrophic for all higher forms of life.

II. The Down Grade Processes.

1st stage: Complex nitrogen compounds to ammonia

The micro-organisms that decompose organic matter keep only part of it in their bodies: much of the carbon is converted into carbon dioxide while some of the nitrogen which does not enter the cells is set free as ammonia. Similarly some of the nitrogen in compounds which they have taken up as food and then digested is excreted as ammonia. The soil insects described later probably excrete their nitrogen as uric acid and some of the soil animals excrete it as urea, but these compounds are rapidly broken down by bacteria with formation of ammonia. However the soil organic matter is decomposed its nitrogen ultimately appears as ammonia.

As an intermediate stage, however, some of it is retained in the bodies of the soil organisms and in compounds which they have synthesised, and the more numerous these organisms are the higher will be the proportion. An easily decomposable protein like dried blood contains far more nitrogen than is needed by the organisms attacking it; some 80 per cent is changed to ammonia hence its high fertilizer value. But if carbonaceous food is added at the same time, such as sugar, starch or cellulose, the organisms can multiply rapidly and in so doing fix so much of the nitrogen in their bodies that a far smaller proportion is left as ammonia. The retention is only temporary: when the organisms die they are decomposed by others and any unwanted nitrogen finally appears as ammonia—but so slowly that it may have little value as fertilizer.

The ammonia so formed is taken up by some of the clay minerals and humus components and held as an exchangeable base; it is held firmly and is not liable to loss either by leaching or volatilisation. But it is rapidly oxidised by bacteria to nitrate.

2nd stage. Nitrification

The first part of the story of nitrification is related on p. 53. Soon after the process was found to be bacterial attempts were made to isolate the organisms concerned but for long they failed. Warington at Rothamsted spent ten years at the task using the best technique of the time: toiling patiently but without success. It was a great disappointment made all the more bitter when Winogradsky, a much younger man, took up the problem and speedily solved it. He realised that the organisms were deriving their energy from the oxidation of ammonia and therefore had no need of organic matter: they could get their carbon from carbon dioxide. He therefore poured his suspension of mixed soil organisms on to a plate of silica jelly containing mineral nutrients and some sulphate of ammonia but no organic matter; only those organisms that could do without organic matter were able to develop. From these plates he was able to pick off colonies of the nitrifying organisms.

Warington had in his meticulously careful experiments been able to show that the reaction proceded in two stages: the ammonia was first oxidised to nitrite, and this was then further oxidised to nitrate, each stage requiring a different organism. Winogradsky confirmed this by finding both sets of organisms on his plates. He was able also to explain the failure of his predecessors: in accordance with the best practice of the time they had all included peptone or some similar organic substance in their media as food for the organisms, and these compounds were not only unnecessary but positively harmful to the nitrifiers. It was one of those tragedies that have repeatedly happened in the history of science. But the search if faithfully done and with scrupulous honesty may not have been in vain. Long after Warington was dead his experimental results carefully recorded in his note books were found to be very helpful in the solution of an important bacteriological problem at Rothamsted.

Only a few organisms can oxidise ammonia to nitrite; and fewer still can oxidise nitrite to nitrate. They all need oxygen, and they do not tolerate much acidity. Minute amounts of iron are needed, but surprisingly enough phosphate does not seem to be necessary in any quantity, at least for some of the strains.[1] The harmful effect of organic substances in the culture medium is not shown by humus in the soil provided it is not acid.

Nitrification proceeds so rapidly in normal soils that neither ammonia nor nitrite occurs in any significant amount, about 2 parts per million of ammonia and even less of nitrite being all that can be detected. This indeed is an extremely fortunate circumstance, as both substances, especially nitrite, are poisonous to plants,—though am-

[1] *Rothamsted Annual Rpt.* 1951. p. 57.

monia in small quantities can be taken up by some—and this earth would have had a very different history if the rate of oxidation of nitrite to nitrate had been slower than that of ammonia to nitrite, instead of being more rapid. For then nitrite would have accumulated, our present vegetation could not have evolved and something radically different would have taken its place.

The oxidation of nitrite to nitrate is a relatively simple reaction, but the oxidation of ammonia to nitrite in the soil is more complex and is not yet understood. Apparently only the exchangeable ammonia is oxidised, but nitrite is not formed at one step. There is some intermediate stage, resulting in a gap between the disappearance of ammonia and the emergence of the nitrite that cannot yet be explained. The obvious chemical explanation is that hydroxylamine and hyponitrous acid are the intermediates but no good supporting evidence has been obtained. Whatever the process it does not involve any loss of ammonia, for the nitrate finally produced corresponds in amount closely with the initial quantity of ammonia.

Nitrification is one of the most important processes in Nature, and the nitrate produced is one of Nature's key products of the highest significance. Without nitrates none of our customary cereal or root crops could grow; there would be some grass and some marsh and bog plants and in the tropics rice, but wheat, the nutritional basis of our civilisation, and most of the plants that have made our gardens and our countryside what they are, would be absent. And, as with nitrogen fixation, this vitally indispensible process can be effected only by very few and very lowly organisms.

3rd Stage. Losses of nitrate from the soil

In spite of its importance, nitrate is one of the most fleeting of all soil constituents. Unlike the phosphate ion which is completely held by the soil colloids the nitrate ion is not held at all but is easily washed out. Also it is readily taken up by plants and by many of the soil organisms. When the air supply is cut off some of the organisms can take their oxygen from the nitrate reducing it to nitrite or ammonia, which may remain in the soil, or to oxides of nitrogen or even to gaseous nitrogen both of which escape into the air. When the air supply is abundant and much organic matter is present in the soil some of the organisms oxidise it by means of the oxygen of the nitrate, liberating nitrogen which is lost into the air. These decomposition processes become more pronounced as the accumulation of nitrogenous organic matter increases, and ultimately they set a limit beyond which it cannot go.

All these changes are reflected in the soil composition as will be shown later.

SOIL FUNGI

The soil fungi form part of a large and widespread family that includes moulds, mildews, mushrooms and other forms. They start as spores, just as bacteria do, but on germinating they produce a white or colourless hollow thread, straight or branched, called a hypha which, growing and lengthening, forms the mycelium. This is the business part of the organism; it secretes the enzymes that attack the organic matter in the soil, especially the plant residues, producing food for the organism. In due course another kind of hollow thread grows out from the mycelium; this produces a fruiting head in which spores develop; later the spores ripen and are dispersed. The white fluffy material on mouldy bread is mycelium, the black pinheads are the spores.

It is only comparatively recently that a soil fungus flora has been recognised. The presence of fungus spores in the soil has long been known, but it was supposed that they were there by accident, having been blown on to the surface and washed down by rain, or carried down by earthworms or other animals. Some forty years ago S. A. Waksman of New Jersey examined numbers of recorded observations and showed that a group of certain genera reappeared consistently in the soils of widely separated countries, though the actual species differed. These he called the "soil inhabitants"; they required oxygen and fed on dead plant remains: in technical terms they were aerobic and saprophytic. Others, more specialised, were parasitic on living plants, some of them only on one kind of plant; as these occurred only in soils in which the host plant was growing they were called "invaders". Their spores, like those of other fungi, could remain quiescent for long periods in the soils but in prolonged absence of the host plant they would die.

The regular soil inhabitants belong to three main groups, the Phycomycetes, the Ascomycetes and the Hymenomycetes. They are aerobic saprophytes feeding on the carbohydrates and the proteins in the plant residues; some among them can decompose and utilise resistant substances like hard cellulose and lignins. Fungi are among the most economical feeders known and can retain in their bodies some 30 to 50 per cent of the carbon they have transformed.[1] With these retained elements they build up their complex cell constituents including chitin, the outer casing of the mycelium, which is of special interest because insects also build up this same substance to make their hard outer skins.

The quantities of fungi in the soil are estimated by methods similar to those used for bacteria: direct counts under the microscope, and cultures on plates of nutrient media. In one of the simpler methods of making direct counts microscope slides or cover slips are buried in the soil or pressed on to a soil surface for a period, the adhering organisms are fixed and stained, the slide is mounted under the microscope and the spores are counted. It is possible also to count the pieces of mycelium but the figures would have little meaning: the mycelium is very fragile and during the laboratory manipulation a single piece may break up into many fragments. The total length of the mycelium is therefore measured, and this, together with the number of spores gives a useful measure of the size of the fungal population.

Unfortunately it is not possible to identify all the spores or pieces of mycelium. This can be done by the culture plate method, but this has two disadvantages: no medium suits all fungi, and some fail to develop; also it is not possible to distinguish between colonies that arise from a single spore or piece of mycelium and those arising from a cluster; the numbers given by this method are therefore too low, probably much too low. But the ability to identify the organisms greatly offsets these disadvantages. S. D. Garrett has neatly summed up the difference between the two procedures: by the direct method you see what you cannot identify; by the plate method you identify what you cannot see.

A good deal has, however, been learned about the mode of life of the soil fungi. As they require organic matter they are more numerous in the surface soil, where more is present, than in the subsoil; they become more abundant as the amount of organic matter in the soil increases. Fertile soils may contain up to about 100 metres of mycelium per gram—nearly 1¾ miles per ounce. In general they are more

[1] For bacteria the figure is more like 1 per cent.

tolerant of acidity than bacteria and there is evidence that they are more abundant in acid than in neutral soils. They develop particularly freely in soils that have been heated: masses of mycelium may form binding the soil into lumps. As the mycelium develops in the soil it periodically sends up into the larger air spaces fruiting heads which there form spores. (Plate V*a*, p. 98).

The soil fungi play an important part in making humus; indeed dead and partially decomposed mycelium seems to constitute no small fraction of the soil organic matter. They play a more indirect part in making plant food. While living they build up their mycelium from the proteins and carbohydrates of the plant residues, and they also compete with the plant for nutrients in the soil thus concentrating nitrogen compounds and other nutrients in their mycelium. When they die this mycelium is attacked by bacteria, the nitrogen is converted into nitrate, and the other nutrients set free.

The process is beautifully seen in the fairy rings, those remarkable green circles that sometimes appear on poor grass land. The ring starts as a green patch, developing where fungus spores have germinated and produced mycelium which assimilates the nitrogen it has derived from the soil organic matter and which had been out of reach of the plant; it thus becomes a reservoir of potential plant food. The patch grows outwards but is not long-lived: the mycelium at the centre dies and is decomposed by bacteria with formation of nitrates and other nutrients that cause the grass to grow better and greener. The grass is removed by grazing animals or by the mower in search of hay; with it goes the plant food stored up by the fungus: the soil, already poor, is still further impoverished. Meanwhile the outer fringe of live mycelium has spread a little further outwards; the inner fringe dies and is as before decomposed with formation of ammonia and nitrate; so a ring of better fed and greener grass appears. This green band is usually about a foot wide: the ring spreads like a ripple when a stone is thrown into a pond, but much more slowly; some are said to have taken many years to reach their present position. The changes were studied long ago at Rothamsted: impoverishment is vividly shown by chemical analysis of the soil.

The rate of loss, if it extended over an acre of land, would amount to 11,500 lb. of carbon and 740 lb. of nitrogen corresponding to about 10 tons of organic matter. The carbon is lost as carbon dioxide, and the nitrogen probably as nitrate, some of which, probably even most, is taken up by the plants giving them their deep green colour; the rest is lost by drainage. The analyses show how destructive are soil fungi of soil organic matter. It is desirable that similar analyses should be made of other fairy rings in this country to see how far these figures are typical. Whatever their actual magnitude the fact of the loss is clear. Some other action appears to be involved: a plant hormone, presumably made by the fungus, has been found in the ring.

TABLE 6

Loss of carbon and nitrogen caused by the fairy ring fungus at Rothamsted.

	UNINFESTED SOIL OUTSIDE THE RING	INFESTED SOIL UNDER THE RING	AFTER INFESTATION INSIDE THE RING
Carbon: per cent	3·30	2·88	2·78
lb. per acre, top 9 inches ..	74,000	65,000	62,500
Loss, lb. per acre	—	9,000	11,500
Nitrogen per cent	0·281	0·266	0·247
lb. per acre top 9 inches ..	6,300	6,000	5,560
Loss lb. per acre	—	300	740

THE PARASITIC FUNGI

The fungi described in the preceding paragraphs are obligate saprophytes; they live on dead plant or animal remains and have no other mode of existence. But there are other saprophytic fungi that can in favourable conditions also make their way into young or weak plants and feed on their juices: these activities are limited because the plant can in some degree protect itself. Others are more potent and can to a greater or less extent overcome the resistance of the plant while still able to live on organic matter in the soil if necessary. Finally there is a group so highly specialised that its members have lost the power of fending for themselves in the soil and can live only in living plants: these are the obligate parasites. They force a way through the outer cortex of the root either by a mechanical squeeze or by secreting an enzyme that decomposes the cell walls. Usually they can attack only one kind of plant or a group of closely allied plants; they are highly specialised and are saved from extinction only by their ability to form resistant spores or sclerotia capable of resting for years completely inert in the soil waiting for their proper host plant.

S. D. Garrett, formerly of Rothamsted and now at Cambridge, has made exhaustive studies of these parasitic forms[1] and groups them as follows:—

1. Soil saprophytes: obligate, no power of parasitising living plants.
2. Primitive parasites: destroy seedlings and juvenile tissues; restricted by the mature host; can also live saprophytically.
3. Less primitive parasites: rapidly destroy plant tissues, less restricted by the host than group 2.
4. Specialised parasites: causing less disorganisation of the host tissues than group 3. No clear evidence of independent saprophytic life apart from survival on dead infected host tissues.
5. Ecologically obligate parasites: complete symbiosis.[2] No destruction of tissues. The organisms have no power of independent existence, and can survive only in the tissues of living host plants; no culture medium is yet known on which they can grow. These are called mycorrhizal fungi.[3]

In discussing the evolutionary significance of this sequence Garratt suggests that the ancestral organisms were saprophytic, but severe competition for food drove some of them to a measure of parasitism on living plants. Gradually this became more specific involving more and more loss of the power to decompose cellulose and lignin, till finally the stage of complete dependence on the living plant was reached.

The fungi causing damping-off of seedlings — species of Pythium and Rhizoctonia—fall into the first group of parasites; they can invade the tender roots of seedlings but not the harder roots of more adult plants. They are greatly favoured by the warmth, moisture, and good soil conditions provided for seedlings in boxes in glass houses, and it is there they do the maximum damage; they are much less destructive in the more austere conditions of the open air. Some plants have protection against these primitive parasites: they can seal off any that have succeeded in getting in. This property varies with different varieties of the same plant: it is one of the factors conferring disease

[1] Summarised in his interesting book *Biology of root infecting fungi,* Cambridge Univ. Press, 1956.
[2] Symbiosis is the living together of two organisms for their mutual advantage.
[3] From μυκης a fungus, and ρίζα a root.

resistance and fortunately it is hereditary. Mary D. Glynne at Rothamsted showed that this explained resistance to Wart Disease of potatoes. Numbers of weak parasites can be found on or even in the root systems of apparently healthy plants, especially those that are weak or wounded.

"Take-all" disease of wheat, which causes serious losses in Canada and Australia, and locally also in England, is caused by one of the less primitive and more specialised soil parasites, *Ophiobolus graminis* much studied by Garrett. It attacks only wheat, barley and a few wild grasses including Agropyron, Agrostis and Holcus. It can live saprophytically in arable soil, but only on residues of its host plants: it dies when their stubble has decomposed or when the available nitrogen is exhausted unless there are other host plants to carry it on. The disease can therefore be avoided by a suitable rotation and system of soil management.

Eye-spot of wheat, caused by *Cercosporella herpotrichoides*, much studied by Mary D. Glynne, is another parasitic fungus that cannot survive without its host plant, living or dead, and therefore can be eliminated by a proper rotation of crops.[1]

One of the most interesting of the soil fungi is *Plasmodiophora brassicae*, an obligate parasite which causes club-root or finger-and-toe in turnips, cabbage, sprouts and other members of the *Brassica* group. Its small round spores on germination liberate myxamoebae,[2] organisms consisting of a body and tail or flagellum that enables them to move to the root, which they enter, lose their tails, cause at first a considerable stimulation of local growth and at the same time they develop a slimy mass called a plasmodium in the root: this finally resolves into spores.

The mycorrhizal fungi are mostly so specialised, and so closely associated with the living plant, that they cannot survive in active form without it. Some are beneficial or even essential to the plant: none are harmful. Forest and heath soils often carry so much litter—dead leaves, twigs, etc.—and are so poor in nutrients, that many seedling trees fare badly or even fail unless the proper mycorrhizal

[1] The Rothamsted Rept. for 1952 (p. 90) gives the following example of the effectiveness of a proper rotation:

Cropping.					Incidence of disease, per cent	
					Eye spot.	Take all.
Wheat following wheat	..	..	..		77	17
„ „ rye grass	..	..	..		24	1
„ „ potatoes	..	..	..		1	Nil.

In both the latter cases two years elapsed since the previous wheat crops.

[2] The prefix "myx" ($\mu\nu\xi a$, mucus or phelgm) implies a slimy mass.

fungus is there to help them. Its mycelium surrounds the rootlet with a sheath from which hyphae force their way into the plant and root-like extensions strike into the soil; there they decompose some of the organic and mineral matter, and pass the products, especially the nitrogen compounds and mineral nutrients, in a dissolved state into the plant. In return the mycorrhizal fungus obtains supplies of sugars and other nutrients from the plant. The association is mutually bene-ficial because the roots of the mycorrhizal fungus are better able than the plant roots to extract the nutrients from the soil and they are also less liable to certain infections; the fungus in turn is assured of a con-tinuous supply of sugar.

The mutual dependence is so complete that soils in forest nurseries may have to be inoculated with the appropriate mycorrhizal fungus if it is not already there. The fungus can continue to survive on the roots even after the tree is dead; the regeneration of birch on moorland soil proceeds better if the stumps of the dead trees are left. This group of sheath-forming beneficial mycorrhizal fungi is called ectotrophic.

Another group of mycorrhizal fungi, the endotrophic, act different-ly. They form no sheath round the root but enter direct, dissolving the cell walls if need be and not necessarily benefitting the plant: indeed Garrett describes the relation as a controlled parasitism rather than a symbiosis. Some, however, are essential. Heath (*Erica*), heather (*Calluna*), bilberries (*Vaccinium*), and other moorland plants need mycorrhizal fungi for successful growth, and as with other plants each has its own particular strain of the fungus. Heaths and heathers introduced into a garden where none have grown before may fail if their appropriate fungi have not accompanied them. The relation of fungus to plant is, however, different from that for trees. Instead of stopping in the root (as in trees) the mycelium spreads throughout the whole plant and gets on to the seed coat; in the case of vaccinium it actually enters the seed; an obviously advantageous arrangement as it ensures the presence of the fungus when the young plant wants it.

Some orchids are dependent in natural conditions on a mycorrhizal fungus—a Rhizoctonia—during their seedling stage. Their seeds are so small that the food supplies stored therein would be quite inadequate to carry the young plant through to the time when their leaves are sufficiently developed to produce their food by photosynthesis. In view of the close adaptation needed the marvel is that the appropriate fungus should ever appear in Nature at the time and place where it is needed. The numbers of seeds are very large and the dispersal fairly wide, but the mortality must be colossal. In horticultural practice cultural treatment replaces the fungus.

Numerous other plants have or can have root fungi associated with them, but it is not always clear whether the association is beneficial to the plant or not. Wheat and clover are among the number; they have been studied in detail at Rothamsted but neither necessity nor advantage could be discovered.

As already stated the spores of some of the fungi causing root disease can remain in the soil for a long time in a state of suspended animation and of course with no power of movement to the roots of a host plant when it appears. The plant, however, excretes from its roots a substance which stimulates the spores into activity: mycelium is put out which, if it can reach the root of its proper host, will enter and continue its life cycle. But the stimulating substance can in some instances be excreted by plants that do not serve as hosts for the fungus and will not receive it; in that case the mycelium, being unable to enter the root, dies and the fungus is finished. This has suggested a method of controlling root diseases. Non-susceptible plants that excrete the stimulant are grown, and the spores germinate; the mycelium cannot find a root that it can enter and so perishes. Rye grass stimulates germination of the spores of the club root fungus (p. 76) but will not accept the mycelium which therefore dies. In experiments in infested soil at Rothamsted 57 per cent of the turnips were attacked after a fallow which did not disturb the spores, but only 19 per cent after a crop of rye grass. The quantity of stimulant excreted is so minute that the discovery of its chemical nature is extremely difficult. Approaching the problem in another way it was found that small amounts of allyl isothiocyanate stimulated the germination of these particular spores just as does the root excretions, and indeed it may be the root excretion.

Another method of dealing with soil-borne disease fungi by subjecting them to toxic secretions from fungi and actinomycetes is described later (p. 89).

THE ACTINOMYCETES

The soil actinomycetes resemble the soil fungi in their general way of life, but they are much smaller, their spores being comparable in size with those of bacteria and so like them that it is impossible to distinguish them under the microscope; direct counts therefore have to include both. But differences appear when the spores germinate. Unlike bacteria they send out fine colourless mycelial threads just as the fungi do, and as in the case of the fungi this mycelium is the active stage in the life of the organism. The commonest species in the soil are the streptomyces; most of these require air and feed on dead plant remains, and about half of those tested produce antibiotics (p. 89). Estimates of their numbers in the soil are subject to the same limitations as in the case of the fungi with the added difficulty that the mycelium is particularly fragile and easily breaks into small pieces. E. A. Skinner at Rothamsted has ingeniously evaded this difficulty and has found as many as 10 to 15 million spores per gram of soil at Rothamsted and elsewhere. The spores themselves are inert but as their numbers in the soil vary at different times it is evident that they germinate and produce mycelium, which in turn produces more spores. It is not yet possible to estimate the total quantity of mycelium with any certainty. Unlike most fungi the actinomycetes are not very tolerant of acid conditions, and they do not occur in acid peats. Some of the actinomycetes are parasitic on plants; among them is the one producing the well known potato scab.

There has been much discussion as to what the actinomycetes are: whether higher bacteria, minute fungi, ancestral types of both, or forms intermediate between them, but it is not necessary to go into this.

The Algae

Soil left undisturbed in mild moist weather is apt to carry patches of green rather slimy growth, especially if it has become rather compact. The growth is a mass of mosses and algae; the latter are small organisms, single cells or filaments, but possessing the green colouring matter which links them with higher plants and enables them to take up carbon dioxide from the air, combine it with water, and form sugars and other carbohydrates using sunlight as their source of energy. Some of the blue green species, especially the Nostocs, can also fix nitrogen from the air, building it up into the protein required for growth. They need the mineral elements essential for all plants, and the nitrogen-fixers require a minute quantity of molybdenum just as do azotobacter and the nodule organisms. A number of different algae are found in the soil: they are smaller and simpler than the more familiar kinds that live in ponds; some, however, appear to be dwarf varieties of these larger species. They have roughly the size of a sphere 10 μ diameter.

The algae are not confined to the surface of the soil. Many are found in the darker depths where no sunlight can reach them and where, if they are living and active, they must obtain food and energy from the soil organic matter like the actimonycetes or fungi. Dr. Bristol Roach at Rothamsted found more algae at a depth of four inches than on the surface of some of the plots, and also more on land well supplied with farmyard manure than on poor land. In spite of many investigations it is not yet known whether they are active in the depths of the soil or whether their spores have simply been washed down by the rain or worked in by cultivation. In the laboratory they will grow in darkness provided they are supplied with nutrients and with sugar; they could find suitable substances in the soil and may be presumed to do so but there is no clear evidence. The green algae seem to predominate in acid soils and the blue green on neutral and fertile soils.

Diatoms differ from other algae by having a silicious coat; they are apparently about as numerous as the green or blue green algae.

Algae in the Rothamsted soil were estimated at 100,000 to 200,000 per gram but much higher numbers have been recorded elsewhere: some 800,000 in Utah and Hungary and 3 million in Denmark. But it still remains uncertain whether they play any part in the depths of the soil.

Their life on the surface however, is in some circumstances extremely important, especially in warm climates. Those that can build up their body substance from carbon dioxide and nitrogen requiring only small quantities of mineral substances from the soil are among the first colonisers of bare ground too remote from growing plants to receive their seeds: places like Krakatoa, devastated in 1883 by a frightful volcanic eruption, or some of the burnt or sterile heaths in England.

In the paddy fields of the east the algae appear to play an important part. The rice is grown in swamp conditions, and the algae copiously growing on the surface of the soil are considered to supply the plant roots with oxygen derived from carbon dioxide and with nitrogen compounds obtained by direct fixation. Further knowledge of this important action is very desirable, but it fully accords with what is known of the algae.

LICHENS

Lichens are made up of algae and certain fungi living together to their mutual advantage, the algae furnishing carbohydrates and where necessary nitrogen compounds, and the fungi providing the mineral nutrients. The lichens of importance for the soil do not live in the soil but on solid rock surfaces or stone walls, on which they often form patches of beautiful golden colour. They are the first vegetable colonisers. The fungal hyphae force a way into the stone so securing a firm hold and also effecting some disintegration and sticking to the detached particles; the rough surface of the lichen catches and holds atmospheric dust. When the lichen dies this mixture of organic and mineral matter forms a basis on which mosses can grow. These continue the accumulation, and in time herbaceous plants can establish themselves.

MICROPREDATORS IN THE SOIL: THE SOIL PROTOZOA

Protozoa are minute unicellular animals formerly supposed to live only in water or very wet conditions: the discovery by E. J. Russell and H. B. Hutchinson that they were important members of the soil population was at first much controverted. Systematic search, however, showed that they were universally present in all soils examined. Some 250 species were found, of which a few occurred only in soil; their distribution was world-wide, and there were no consistent differences in the fauna of soils from arctic, temperate and tropical regions, nor did differences in soil conditions seem to produce much effect. In this respect they resemble the fungi. They occur in two states in the soil: an active state, and an inert or resting state called a cyst, which is surrounded by a cell wall and can withstand prolonged drought or absence of suitable food. Cyst formation, however, is not always a flight from adversity: for some species it is part of their natural life cycle.

Two main groups of protozoa occur in the soil: flagellates and amoebae. The flagellates are about 3 to 10 μ in length, they have a tail or flagellum (some have two or more) varying in length from about 5 μ to 20 μ which enables them to move. The amoebae are larger

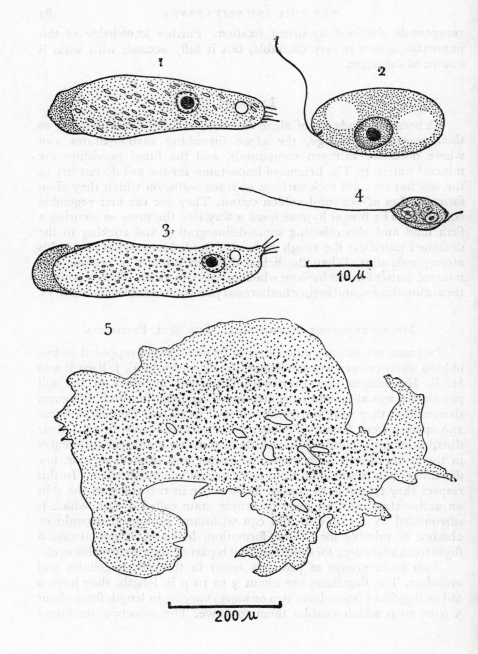

though still small compared with those found in water; they are about
10 μ to 40 μ across, they have no permanent flagella, but being naked
blobs of protoplasm unconfined by a cell wall can change their shape
and so propel themselves. There are many kinds of them in the soil,
and improvements in methods of search steadily reveal more. (Fig. 5,
opposite. One group of amoebae is covered with a hard shell. A few
ciliates are present but the conditions hardly seem suitable for them.

Multiplication is by division. The cell has a well-defined nucleus;
this breaks into two parts which separate: protoplasm gathers round
each and two organisms appear in place of one. The process is actually
very complicated. Different species of amoebae differ in the ways in
which their nuclei divide, and it is on that basis that they are classified.
The flagellates have rather wide nutritional possibilities: they can feed
saprophytically like fungi and indeed one, *Euglena,* has green colouring
matter and can live like an alga; chiefly, however, they feed on bacteria.
Amoebae also feed chiefly on bacteria which they capture, surround
with protoplasm and digest: other small organisms—algae, protozoa,
yeasts—can however serve as food: one amoeba even feeds on the
potato eel worm. Not all bacteria are equally acceptable; some are
readily and completely consumed; others, including many of those
producing blue, green, red or fluorescent pigments, are not taken,
some indeed appear to secrete substances poisonous to the amoebae.
B. N. Singh, to whom much of the recent knowledge of soil protozoa is
due, offered 87 strains of bacteria to eight predators, each accepted
about half of the strains but not the same half; only 12 were acceptable
to all the predators: and only seven were rejected by all. Azotobacter
seem to be specially appreciated, but the nodule organisms apparently
were not. Among those accepted there are differences in nutritive
value as shown by the differences in size of the amoebae and of their
rate of multiplication. Some kinds of amoebae will not come out of
their cysts unless the appropriate bacteria are present.

Methods have been devised for estimating their numbers in the

---

← Fig. 5 (opposite)
Soil protozoa: 1. *Didascalus thorntoni* (amoeba); 2. *Oikomonas termo* (flagellate);
3. *Schizopyrenus russelli* (amoeba); 4. *Cercomonas crassicauda* (flagellate). All magnified
1,500 times; 5. *Leptomyxa reticula* (Giant Rhizopod.) Magnified 150 times.
(Drawn by B. N. Singh). For descriptions see his papers in *Phil. Trans.* (1952), B.
236, 405-461, and *J. gen. Microbiol.* (1948), 2, 59-96.

soil; they appear to fluctuate considerably even from day to day. In a striking pioneering study D. W. Cutler, Lettice M. Crump and H. Sandon showed that the numbers of the most prominent of the amoebae on one of the Rothamsted plots fluctuated daily in inverse ratio to the numbers of bacteria as given by the plate method. Later work confirmed this, but showed that the total numbers of bacteria did not vary in this way: suggesting that the plate method, while recording bacteria preferred by the amoebae, failed to record all that were rejected.

At Rothamsted the amoebae, like the bacteria, are most numerous on the plots receiving farmyard manure; they are more numerous on the plots receiving complete fertilizer than on those unfertilized while the bacteria are not—which suggests a difference in nutritive value of the bacteria. The protozoa keep down the numbers of the bacteria but they do not impair the efficiency of the survivors.

The Total Quantity of Micro-organisms in the Soil

Owing to the great difference in size of the different members of the soil population a mere comparison of their numbers gives little information about their respective quantities. Unfortunately no precise figures for the weights of the different members can be obtained but an estimate has been made by E. W. Russell based on certain defensible assumptions and which though subject to a considerable margin of error shows the orders of magnitude:—

Weights of organisms in top 6 ins. of Rothamsted soil
(E. W. Russell)

	lb. per acre
Bacteria	1500—3500
Fungal mycelium	1500
Flagellates and amoebae ..	150.

This corresponds to about 2 or 3 per cent of the total organic matter of the soil. The list is incomplete but no estimate could be made of the weights of the missing organisms.[1]

Other Predators: Giant Rhizopods, Acrasieae, and Myxobacteria

By improving the methods of searching for micropredators B. N. Singh was able to show that giant rhizopods, formerly thought to be

[1] Some of the Continental estimates are higher (p. 135).

only rare inhabitants of the soil, are really wide spread in arable soils, though not in grassland. They are extraordinary organisms and may be as much as 3 mm. in diameter, but being spread over the soil particles cannot be detected till they are removed and grown in a culture medium. Like the amoebae they feed only on selected bacteria, and rejected more than half of the 92 strains Dr. Singh offered them; their choice was not the same as that of the soil amoebae, but like them, they refused the pigment formers. They are more catholic in their selection than amoebae and at least one species will also feed on the eelworm that attacks potatoes. Numbers exceeding 1,000 per gram of soil have been recorded at Rothamsted. (Fig. 5, p. 82).

Another remarkable group of organisms revealed by these improved methods of search are the acrasieae which have something of the nature of amoebae and also of fungi. In the active state they look like amoebae, and further resemble them in feeding on selected bacteria and reproducing by simple fission; but they can also collect together and form fruiting bodies that look like those produced by fungi. Inside these, spores are formed from which when they burst a new generation of amoeboids emerge. These are called myxamoebae. These organisms appear to be more numerous in arable than in grassland soils.

Other soil organisms, the myxococci group of myxobacteria, first kill their bacteria by a toxic excretion, and then liquify the bodies by an excretion and absorb the products. These also attack only selected bacteria. Like other myxobacteria the myxococci were formerly supposed to live only in dung, but improved methods of search have shown them to be present on all the Rothamsted plots examined, even those that have received no dung for a century or more. Some 2,000 to 76,000 per gram of soil have been found on the different plots on various occasions.

This does not end the tale of predacity in the soil, for some of the fungi capture and feed on amoebae and nematodes, and some of the nematodes and other small animals feed on bacteria.

The Bacteriophages, often simply called Phages

The phages are the most mysterious of all the agents capable of destroying soil organisms. The name is non-committal and simply means a devourer. They are viruses, and as such are unlike any living organism or any dead inert matter: they have some of the attributes of both. Their presence in soils was first shown by a French scientist, F. d'Herelle, in 1915 but little could be learned about them because suitable appliances had not yet been invented. The difficulty is their minute size: no optical microscope can reveal them. It was not till the electron microscope was developed, with a resolving power some 600 times that of the best optical microscope,[1] that much progress could be made with the study of phages.

The particles are almost spherical: some are only about 0·01 μ in diameter, i.e. a hundred times smaller than the smallest of the bacteria: others are larger, ranging up to about 0·1 μ in diameter. Usually but not always there is an appendage which may be as much as 0·3 μ in length and about 0·01 to 0·025 μ in width. The spherical part is commonly called the head, and the appendage the tail. (Plate Vb, p. 98).

The head contains a nucleic acid (deoxyribonucleic acid)[2] surrounded by a membrane of protein, but nothing in the nature of a nucleus has been seen. The tail consists of protein.

Phages are specific in their action: a particular phage can attack only certain strains of one group of bacteria. For almost all bacteria and for some actinomyces, however, there is an appropriate phage. The phage attaches itself to the bacterium by means of its tail, and through this the contents of the head are forced into the bacterium. It then completely disappears leaving no visible trace behind. This is called the stage of eclipse. Although nothing can be seen something is obviously happening, for after a time new phage particles appear in increasing numbers in the bacterial cell; the numbers vary: there may be hundreds of them; finally the cell dissolves and the particles are liberated. The time elapsing between infection and dissolution of the bacterial cell with liberation of the phage varies with the phage; usually about half is occupied with the stage of eclipse.

[1] A good class optical microscope can magnify about 1000 times and can see 1 micron (μ) or 10,000 Å (p. 10). Present day electron microscopes can see 15 to 20 Å and will probably soon get down to 10 Å. But they are expensive; a recent estimate for a good instrument was £11,000. The chief experimental difficulty with biological material is to get it into a form suitable for examination by these instruments.

[2] The presence of deoxyribonucleic acid is interesting because this acid is one of the key substances of life: it may indeed be the medium by which hereditary characteristics are transmitted from generation to generation. Its molecule is large and very long; it consists in two intertwined spiral chains each built up of sugar molecules on the outside of which are phosphate groups and on the inside various bases linked to each other across the axis of the molecule. There may be a thousand or more turns of the helix in the molecule.

This complete solution of the cell is called lysis. On it has been based a simple method of estimating the number of phage "particles" in a given weight of a soil or volume of a liquid culture. A suspension of phage from a known quantity of soil is mixed with a culture of susceptible bacteria and poured on to a plate of nutritive material; the bacteria develop making a visible background of continuous colonies on which clear spaces made by the phage particles stand out and can be counted.

The bacteria are not without defence: in presence of the phage they can develop resistant strains. These do not all retain their resistance when grown apart from the phage, and in multiplying will produce a certain number of non-resistant strains.

The phages attacking Rhizobium, the nodule organism, have been most fully studied; they have been extracted by Janina Kleczkowska from soils at Rothamsted and Woburn on which leguminous crops have been growing for some years. It was at first thought that they were responsible for "clover sickness", the inability of a soil to carry a clover crop for more than a few years in succession. No proof of this has been obtained; inoculation of healthy soils by phages has failed to produce the effects of clover sickness; nevertheless it remains true that phage is most easily extracted from soils on which clover has been growing for several years.

MODE OF LIFE OF THE SOIL POPULATION

Little is known of the mode of life of the soil population. No method has yet been devised for looking into the soil to see what is happening, but an ingenious beginning has been made by two British investigators, F. E. D. Alexander and R. M. Jackson. A block of soil was suitably stained, impregnated with a transparent plastic and cut into sections for microscopic examination as is done in the study of rocks. Not much use has yet been made of this method but in the meantime a surprising amount of knowledge of the soil life has been obtained by indirect methods. The great difficulty is the small scale on which everything is happening in the soil. One tenth of an inch seems very small to us, and it is quite impossible to estimate the difference in climatic or other conditions between one end of the distance and the other. But to one of the soil bacteria the distance is as great as three miles is to a man. Our estimates of soil conditions lump together regions which in relation to the soil organisms are as large as if we used a generalised average of conditions over most of Europe in discussing the effect of the weather of Manchester on the activities of its people.

It is known that the micro-organisms are not evenly distributed throughout the soil. They concentrate in considerable numbers around the roots of growing plants,[1] especially leguminous plants, organisms that require the simpler growth substances especially those containing sulphur, like methenionine, being particularly numerous. It is not known how or why the organisms should collect: they may be fed or merely stimulated to multiply by the root excretions. In the case of leguminous plants like clover and lucerne some of the organisms are directly beneficial to the plant and their multiplication is known to be stimulated by a root excretion (p. 65).

Some plants can, however, excrete substances that protect them against attacks of parasitic fungi. Varieties of flax resistant to Wilt disease owe their immunity to the hydrocyanic acid which their roots excrete and which poisons the Fusarium and Helminthosporium fungi that cause the disease, but stimulates the Trichoderma that also represses these fungi. Susceptible varieties of flax on the other hand do not excrete hydrocyanic acid and their roots become surrounded by a mixed population of fungi including those causing the disease.

The advantage to the plant of being able to excrete or secrete protective substances is obvious; it is difficult to understand why some plants, instead of leaving inert forms at rest, secrete substances that stimulate the germination of spores of parasitic fungi or bring active eelworms out of their inert cysts. The potato is a tragic example: it secretes into the soil substances which quicken out of their dormancy any spores of the destructive wart disease (*Synchytrium endobioticum*) or cysts of the equally destructive eelworm *Heterodera schachtii* that may be present and in due course it succumbs to their attack. Some varieties have the power to seal off the invader before it has progressed far and they are classed as resistant; but many of the most desired varieties cannot do this; they are susceptible. The evolutionary significance of this action in awakening a sleeping enemy that would perish if left undisturbed is an interesting topic for speculation.

One of the most serious factors in the life of the soil population is the competition for food which apparently is intense. The population

[1] This zone is called the rhizosphere and has been much studied by A. G. Lochhead and his colleagues in Canada.

seems to be Malthusian in that it is living right up to the available resources. Fragments of organic matter are crowded with micro-organisms. Fungus mycelium even before it is dead may be severely attacked by bacteria. Addition of organic matter greatly increases the numbers of any groups that can attack it, but when it is decomposed they fall again; even some most unlikely substances can bring about marked increases in the numbers of those organisms that can utilise them.

More positive injury is done by some members of the population. Some become parasites on others. The predators feed upon some, but may, in turn be caused to suffer. Large organisms may deprive smaller and more delicate ones of oxygen and produce volumes of carbon dioxide that asphyxiate them. But there is a more drastic way of antagonism. Many of the fungi, actinomycetes, and bacteria can produce and excrete substances which are highly toxic to other organisms. The action is selective, only certain organisms being killed and the name poisons would thus hardly be appropriate; they therefore receive a special name, antibiotics. The story of these is one of the romances of modern science. It was long ago observed that a parasitic fungus produced more serious infection on plants growing in sterilised soil free from all other organisms than on plants growing in untreated soil containing the normal soil population. Introduction of various soil organisms reduced the potency of the parasite and lists were drawn up showing which were the most effective antagonisers. Among them were bacteria, actinomycetes, fungi especially mushrooms and toad-stools; practically all were aerobic spore-forming saprophytes, and, most important, all could produce and excrete antibiotics. It was presumed, and there is considerable evidence in support, that the antagonism was due to these substances.

The existence of some antagonistic factor had long been known. Some of the colonies of micro-organisms on culture plates were often observed to surround themselves with a clear ring in which nothing would grow. But bacteriologists could get no further till organic chemists had improved their technique so as to be able to isolate and study minute quantities of substances; when this was done rapid progress was made. This is a common experience in scientific investigations: progress often has to wait on advances in some other subject.

The medical people were the first to study antibiotics: penicillin is now widely used to kill harmful organisms that have got into the human system. Their use for plants is investigated at the I. C. I. Research laboratories at Welwyn by P. W. Brian and his colleagues. Tomatoes, lettuce, oats and barley were protected against certain fungi by supplying them with the antibiotic griseofulvin. This will become practicable on the large scale when supplies are available; in the meantime alternative methods have been tried. Fresh plant material has been added to the soil by green manuring: this favours the development of the saprophytic organisms including those producing antibiotics, but it does not favour the parasitic organisms; on the contrary the additional carbon dioxide generated from the plant residues is likely to be harmful to them as also is the additional competition for essential nutrients from the soil consequent on the greatly increased numbers of the saprophytes. This method has been successfully controlling scab in potatoes. The interpretation of the results, however, is by no means simple as various other effects are produced by the added organic matter, including a change in soil reaction.

Another method of which great hopes were entertained, was to introduce into the infested soil an antagonist potent against the parasite that was destroying the crops. In the laboratory experiments the method worked admirably: plants grown in sterilised soil containing the parasite were badly attacked, but trials with other fungi introduced simultaneously revealed a potent antagonist. So long as the parasite and its antagonist were the only organisms present the parasite was properly kept in check. "Take-all" in wheat, damping-off of seedlings, and other troubles caused by soil parasitic fungi were satisfactorily controlled by the fungus *Trichoderma viride* which produces the antibiotic gliotoxin. Unfortunately no such happy result was obtained in natural soil when all the other soil organisms were present. Some bitter disappointments have ensued. The Panama wilt disease has caused great losses of bananas in the Trinidad plantations, and as bananas are the second most important export a remedy was urgently needed. The disease is caused by a soil fungus, and a potent antagonist was found which, so long as it had a clear field in sterilised soil, exerted satisfactory control. But, as happened with Take-all and damping-off, the antagonist was ineffective when added to the soil of diseased plantations.

No satisfactory explanation of this puzzle has yet been found. It has been suggested that the antagonists among the soil organisms (for there are often a number of them of varying degrees of potency) are mutually antagonistic each to the others, and the introduction of one more can have little effect. Some investigators have even doubted whether any antibiotics are produced in the soil: none has yet been extracted from it. Against this is the fact that some at least of the antibiotics are firmly held by the clay colloids and could not be removed although they remain as potent as ever. It is also argued that antibiotics are so easily decomposed by soil bacteria that none could survive for long. This may be so, but it is not necessary that they should survive. The antibiotic does its work quickly, striking down its quarry: the fact that it provides food and energy for bacteria is immaterial in this connection though fortunate from the wider point of view. For if antibiotics could accumulate in the soil the whole sequence of changes would be affected. More sensitive tests have strongly indicated the presence of antibiotics in the soil: the characteristic deformation of the hyphae of certain fungi which they induce has actually been observed.

It is possible to narrow down the soil population by changing the reaction of the soil and so obtaining conditions for successful control. J. Rishbeth at Cambridge showed that the root disease of pines caused by *Fomes annosus* and serious on non-acid soils in East Anglia could be controlled by *Trichoderma viride* when the soil was acidified—a change that eliminated many members of the soil population. It has long been known that the potato scab organism could not function in acid soils: the practical difficulty is to make a neutral or alkaline soil acid.

One might speculate indefinitely as to the value to the organism of this power of producing antibiotics. It is an obvious advantage in the struggle for food to be able to discharge a volley of poison on to competitors. If, however, as some suppose, the antibiotic is not produced in normal conditions it is difficult to see how the power of producing it can have been retained. Nor is it any objection that the amount of organic matter in the soil is only small, and the production of antibiotics requires rich nutrient conditions. The organisms are extremely small, and the particles of organic matter on which they are acting may seem very large to them. The subject is so new and fresh knowledge is perpetually being gained: all that we can do is to wait patiently for further advances.

Whatever life in the soil is like it certainly is not a placid system of organisms working together harmoniously for the cleansing of the soil from unwanted organic matter and the production of nutrients for plants growing for our pleasure or benefit, although these things do come about. So far as can be seen the struggle for existence seems to be at least as intense within the soil as it is on the surface, while the hazards of life are probably greater.

Water extracts of soil not infrequently facilitate the growth of micro-organisms suggesting the presence of something in the nature of growth promoting substances, but none has been actually isolated.

CHAPTER 5

THE SOIL INHABITANTS: II. THE VISIBLE MEMBERS: THE ANIMALS

The Meio- or Meso-fauna

MANY OF the organisms already described are associated with the water in the soil. The animals on the other hand, apart from the protozoa and nematodes, live in the air spaces or in burrows of their own making. There is an amazingly large variety of them; H. M. Morris found 72 different species of insects in one of the Broadbalk wheat plots at Rothamsted, and insects formed only half of the total animal population. The smallest are almost invisible, they include the mites, collembola, and others; the largest are the earthworms.

In order to see this vast and varied population in proper perspective it was necessary to devise means of estimating the numbers of its different members in a given quantity of soil. The principle is simple: the sample of soil is introduced into a cylinder filled with liquid to which more is added from below so that it continually overflows. The soil, being heavier than the liquid, sinks to the bottom while the animals, being lighter, are carried upwards. The overflow is directed on to a series of sieves which keep back everything larger than the particular mesh; animals are picked out and identified. In practice of course there is much refinement: the soil has to be completely but gently disintegrated and the animals separated from its particles; the liquid is generally a salt solution of somewhat higher density than water and therefore better able to float out the animals, and supplementary treatments are adopted to separate insects and mites from scraps of vegetable matter which also float. As the methods improve so the varieties and numbers revealed become higher.

Three different habitats have been studied in England: cultivated soils, grassland, and heath soils; unfortunately the studies were not made at the same time or by the same methods so that the results are not strictly comparable. H. M. Morris in 1922 found a total invertebrate population of 15 millions per acre on the plot on Broadbalk wheatfield receiving annual dressings of farmyard manure, and 5 millions per acre on the unmanured plot. About half were insects. Methods have since been greatly improved, especially for catching the smaller organisms, and later investigators on other sites have obtained far higher numbers[1] K. D. Baweja in 1936 found 68 million invertebrates per acre on an old allotment soil at Rothamsted, 45 million of which were insects; while still later J. M. Jones estimated that the soil of the experimental garden at Aberystwyth contained 160 million arthropods per acre, without reckoning other animals.

The grassland studies were made in 1943 by G. Salt and his colleagues using still better methods. They were confined to the arthropods, but even so the numbers far exceeded those of Jones and Baweja, and it was recognised that they were incomplete because many of the minute forms, larvae, etc. were still not being captured. Samples were taken at two depths: 0—6 inches and 6—12 inches: the figures obtained are recorded in Table 7

TABLE 7

Numbers of arthropods at different depths in grassland. Millions per acre.

	INSECTS				OTHER ARTHROPODS			ALL
INCHES	COLLEM-BOLA	HEMI-PTERA	OTHERS	TOTAL	ACAR-INA	OTHERS	TOTAL	ARTHRO-PODS
0 – 6	174	62	35	271	485	4	489	760
6 – 12	73	10	25	108	181	20	201	308
Total	248	72	60	379	666	24	689	1069

Allowing for the numbers missed the total Arthropod population was estimated at about 1,400 millions per acre of which 950 millions were mites and 280 millions Collembola.

P. W. Murphy in 1953 by a somewhat different method made extensive studies of the fauna in heathland in its natural state, very acid, pH3, with a cover of natural litter, and no earthworms—the forester's "Mor"—and on cultivated heathland planted with trees

[1] A great advance was made when W. R. S. Ladell introduced his salt-flotation method of entracting the minute soil animals.

where there were earthworms. Here far higher numbers were obtained: 2·3 thousand million arthropods per acre on the natural heathland and 3·4 thousand million on the cultivated; this last is the highest figure yet recorded in Great Britain though it is exceeded in Sweden. Here also the most numerous were the Collembola and the Acarina.

The differences in methods make close comparisons of the above results impossible. But it is clear that the population is much denser in undisturbed grassland soils where there is plenty of vegetable matter, dead and alive, as sources of food, than on cultivated land where the supplies are less; the population is denser also where farmyard manure is applied than where it is not.

There is great diversity among these minute animals but they have certain things in common. They live in the air spaces in the soil but cannot alter them; unlike the worms they cannot make burrows for themselves: they have to be content with the *Lebensraum* already existing. Most of them avoid the light: those living near the surface may have eyes, but those living lower down have not; instead however they have sensory bristles (setæ) or other compensating organs. Other adaptations to their habitat are their flattened bodies and short legs.

The smaller animals are much more numerous than the large ones, and P. W. Murphy points out that the weight of the soil fauna in a given weight or volume of soil—the "Biomass"—is a much more useful standard of comparison than mere numbers. He further points out that the small animals consume more oxygen per unit weight than the large ones, showing that they are effecting proportionately more decomposition of organic matter. (Fig. 6, p. 96).

In all the soils studied Acarina and Collembola are by far the most numerous members of the population. The Acarina or mites are found all over the world, from the tropics to the Arctic, and from the desert to the rain forest. A number of species are found in British soils. Their size range is considerable: from 0·1 mm. to 2 mm. The body appears to be unsegmented giving the appearance of a very simple organism; indeed mites have been described as sacs with legs.[1] (Plate VI, p. 99). The range of feeding habits is very wide: some species feed on dead vegetable matter, some on the excreta of other soil animals: these are called coprophagous; others are parasitic on other soil animals or have

[1]Salt estimated the volume of the smallest as something between 5×10^{-7} c.c. and 4×10^{-6} c.c., i.e. some 12,500 to 100,000 could be packed into a drop of water. Certain assumptions involved in the calculations are not accepted by all Soil zoologists.

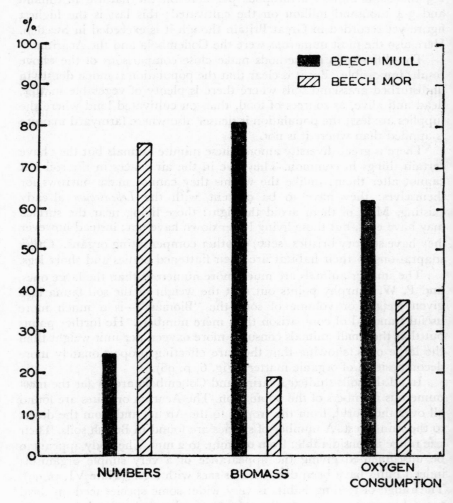

FIG. 6
Relative numbers, weights, (biomass) and oxygen consumption of meio- and
macrofauna in two forest soils, beech mull and mor. Each column represents
the percentage of the total for mull and mor. (P. W. Murphy, in Kevan,
Soil Zoology).

commensal relations with them. The commonest in British soils are tough skinned slow moving creatures probably feeding on decaying vegetable matter, fungal hyphae and spores. Others again are predators feeding on nematodes, small worms, insect larvae. One particularly disagreeable animal is the harvest bug which is very common on grassland in later summer and early autumn: it is really the larval stage and not the adult. It pierces the soft skin of one's legs and sets up severe irritation. It does not however remain permanently in the body, but leaves it to enter its adult stage. The simplest retaliation for the sufferer is to touch the spot with a drop of a potent antiseptic such as Dettol when the mite is at once killed.

Reproduction is in general sexual: from the eggs larvae hatch out having six legs, these pass through two or three nymphal stages with eight legs and finally become adult: the whole cycle may take a year to complete in some species. Reproduction may also be partheno-genetic, i.e., without the intervention of the male. Little is known about their way of life in the soil: observation is extremely difficult.

Collembola or Springtails are perhaps the most interesting of all soil insects because of their great antiquity. They undergo practically no metamorphosis; they are wingless, and seem to be survivors from the days before insects had developed wings: a primeval race left stranded in the march of evolution. An insect closely resembling a present day Springtail has been found in the Lower Devonian rocks in Scotland: this era is some 300 million years ago and some 40 or 50 million years before the earliest winged insects appeared in Upper Carboniferous times. They are segmented grub-looking animals with six legs and a large head provided with eyes and two antennae; often brightly coloured, especially those living near the surface of the soil: the size of the different species ranges from 0·3 mm. to 5 mm. Their characteristic feature is a remarkable organ at the rear end by which they are enabled to jump a foot or more, which in relation to its size is as if a man could jump some 250 or 300 yards. (Plate VI, p. 99). It is, however, only the surface dwellers that possess these attributes: those living lower in the soil lack the bright colours, the eyes and the spring organ of the surface dwellers; instead they are white, eyeless and sluggish, the spring being either very degenerate or altogether absent.

Moisture is very necessary for them, and they can even survive short periods of flood. They are much less specialised in their food habits than the mites and feed on decaying plant residues, fungal hyphae and spores, algae, pollen grains, also on dead animal matter: flies, fly casings, other Collembola and their castings; as in the case of the mites, however, the majority probably feed on dead plant residues and fungi. They have on occasion caused damage to mangold crops at Rothamsted. Reproduction is sexual and more rapid than in the case of the mites: the eggs are laid in batches of six to twelve; nymphs hatch out direct, and unlike most other insects they look like the adults except that they are smaller and colourless. They moult several times before reaching the adult stage. Their numbers are highest when temperatures are around 45°F. to 55°F. and the soil remains moist: they fall considerably at temperatures below 45°.[1]

They have of course their enemies: spiders, beetles, and some of the mites prey upon them. They show a curious tendency to congregate at times.

There are many other kinds of minute animals in the soil, less numerous than the mites and the springtails, but equally interesting and no doubt important in the soil economy. Two deserve special mention: both are very minute and related to Collembola. The Thysanura or Bristle-tails are among the most primitive of all insects. The Protura are also very primitive, lacking eyes, antennae, and like the other two, hardly undergoing any metamorphosis. Some of them are only 0·5 mm. in length and 0·1 mm. in maximum breadth. They were supposed to be rather rare, but search on the Rothamsted grass plots showed that they are actually quite numerous especially on the plots that are not acid: they inhabit the lower six inches of soil rather than the upper layer. (Plate VI, p. 99). It is the old story of better methods revealing unexpected organisms in unsuspected numbers.

The soil aphides also tend to live in the lower six inches of soil. Little however is known about them or their function in the soil. The reader desiring information about them should consult one of the standard text books on entomology.

The Nematodes (Eelworms)

The name means "thread like" and it is well chosen, for until they are magnified soil nematodes look like small pieces of very fine thread.

[1] At high altitudes and in arctic regions Collembola are often very numerous— they are called Snow fleas—and sometimes constitute the entire soil fauna. P. W. Murphy found them in great numbers in Austria at a height of 9,200 ft. in almost barren rock detritus; there were practically no mites.

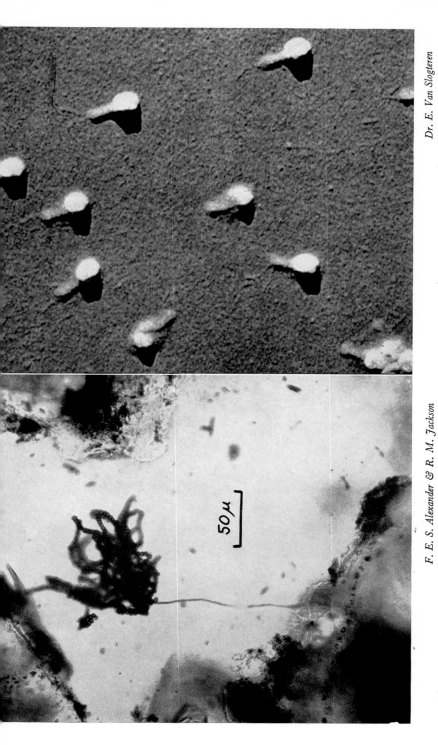

F. E. S. Alexander & R. M. Jackson

Dr. E. Van Slogteren

Plate Va. Conidial head of *Aspergillus niger* in an interstitial space in the soil.

b. Bacteriophage × 56,500 (*Electron micrograph*)

50 μ

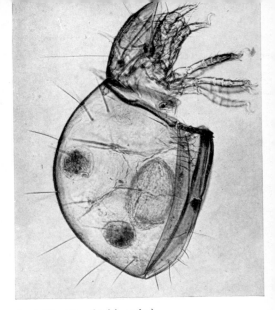

a. Collembola (*Isotoma   sensibilis*)
    × 60

b. Mite (*Pseudotritia ardua*)

d. Mite (*Pergamasus runcatellus*).

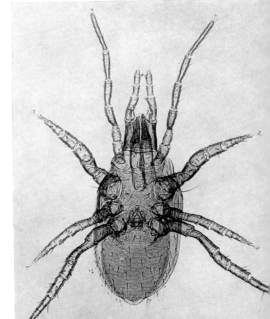

c. Protura × 65

Plate VI. Minute soil animals. (*Photograph a,
Dr. C. A. T. Edwards; b, c, P. W. Murphy &
H. F. Woodward; d, V. Stansfield*) *Jl. Soil Sci.*
(*1953*) *4, 155-193*

They range from about 0·5 to 1·5 mm. in length which means that some 20 to 50 go to the inch; they are unsegmented and mostly very thin, being 40 or 50 times as long as they are broad.

Originally they lived in water and many kinds of them do so still. But some kinds migrated to the land and although their close dependence on water still continues, they find enough in the pores and films on the soil particles to serve their purpose. Some lead a completely independent existence: these are called the free living forms. Others are parasitic on plants and cannot complete their live cycle without them; while others have invaded animals and cause serious diseases: some of these are of extraordinary size, running to many feet in length. Many thousands of species are known. Some hundreds have already been found in the soils of different countries: in the standard treatise on the subject[1] T. Goodey gives drawings of 189 different species of free living forms and new ones are still being discovered. Some 60 species are regarded by Overgaard Nielsen as constituting the typical soil fauna in Denmark but some of these are not numerous and only about 20 are abundant in all kinds of soil. The soil forms vary considerably among themselves: the largest found in the Danish soils was 2,500 times the size of the smallest, and a size difference of 500 times was not uncommon even within the individuals comprising a single sample. These outsizes, however, can be regarded as exceptional and proving only the great variability of the organisms: the usual sizes are those given above.

It is not difficult to get the nematodes out of the soil so as to look at them under the microscope. Some soil freshly gathered from a garden bed or better from a grass field is tied up in a piece of cheese cloth and put into a glass funnel, the stem of which is stopped by a piece of rubber tubing and a pinch cock. The funnel is then filled with water, the nematodes wriggle their way out of the soil and through the cheese cloth, finally collecting in the stem. After a few hours some of the water is run into a Petri dish and search is made for the nematodes; they can be picked up by a fine bamboo splinter and mounted on a glass slide in the usual way.[2]

Examined under a microscope the nematodes are seen to be highly

[1] T. Goodey, *Soil and Fresh Water Nematodes*, (1951), London, Methuen, also *Plant Parasitic Nematodes*, (1933), Methuen.

[2] Full details of the technique are given in *Laboratory methods for work with plant and soil nematodes*, (1956), Tech. Bull. No. 2. 3rd Edn., London, H.M.S.O.

WOS—H

complex in structure. (Plate VII, p. 114). Even the small ones, only one-fiftieth of an inch in length and vastly smaller in girth, have an almost complete set of animal organs. There is a head, which varies considerably in the different species: in some it has six lips, two subdorsal, two subventral, and two lateral; the mouth opens into a cavity—the buccal cavity—which in some species is provided with teeth, in others with grinding surfaces; behind this is the oesophagus corresponding to a throat, which leads to the intestine (there is no special stomach) thence to the rectum and the anus. The oesophagus is muscular and forces the food from the mouth into the intestine where it is digested; the assimilated material is stored in the cells forming the intestine walls and the rejected portions pass into the rectum and are excreted. A pore on the ventral side allows of excretion from a simple system of tubules which run longitudinally almost the full length of the eelworm. There is a nervous system but nothing corresponding to heart or lungs: no circulatory or respiratory system; the body cavity, however, contains a fluid which bathes the various organs and serves both respiratory and nutritional purposes.

Reproduction is in general sexual, there being both male and female forms; the female lays eggs which in due course hatch out producing larvae. Some of the egg-laying performances are almost incredible. Nielsen records that a *Cephalobus* female within 22 days laid five times her own weight of eggs—209 in all. In many species males are not known, however, and the organisms are hermaphrodite, sperms forming in the gonad prior to the development of the ova. In other species reproduction is parthenogenetic. As the larvae grow they have to shed their skins or cuticles: this may happen four times before the adult stage is reached and reproduction can begin. From egg-hatching to egg-laying may take some 20 to 30 days. The shed cuticle may not be completely detached but may remain as protection against adverse conditions.

For active life nematodes must have moisture, but in dry conditions instead of dying many can pass into an inert state in which they can survive for very long periods. They can then readily be picked up with the dust by the wind and carried considerable distances; they can also be carried on the feet of birds or by insects. The commonest species are widely distributed.

Nematodes are more numerous on grassland than on cultivated land: in the former there may be some 20 million per square metre

(1·2 sq. yards) in the latter, half to 2 millions. Dung contains large numbers of them, and temporarily raises their numbers when added to the soil, though they gradually die down to the normal density. They concentrate near the surface: for most species Overgaard Nielsen found the large majority—up to 90 per cent—in the top 2 inches of soil and none below 4 inches. The fauna is substantially the same both in numbers and character whether the soil is a sand or a clay, but an acid humus soil carried a smaller and a different fauna. There is no indication of seasonal variations in number.

The smaller nematodes feed on bacteria, small algae, and actinomycetes; larger ones provided with teeth or grinding plates feed on protozoa, rotifers, and other nematodes. It used to be thought that they fed also on decaying organic matter, but Nielsen examined the gut of large numbers of soil nematodes and could find no evidence that they do so. They certainly abound in decaying organic matter, and also near the roots of plants, but so do bacteria and other micro-organisms, and it may well be that it is these on which they are feeding.

In conditions where the total numbers are small the feeders on algae preponderate; as conditions become more favourable the feeders on bacteria increase considerably and to a less extent the root-browsers also. Nielsen found that as the numbers rose from 175,000 to 20 million per square metre the bacteria-consumers rose from 7 per cent to 47 per cent of the total; the root-browsers from 4 to 16 per cent; but the algae-consumers decreased from 88 to 28 per cent. These latter preponderated in very dry soil, also in fresh water and in acid humus soil, where, however, bacteria are relatively few.

With great ingenuity Nielsen estimated not only the numbers but also the weights and the oxygen consumption per hour of the total free living nematode population on his experimental areas in Denmark. His results were, per square metre:—[1]

		NUMBERS MILLIONS	WEIGHT GRAMS	OXYGEN CONSUMPTION C.C. PER HOUR AT N.T.P.
Grassland	.. ..	4 – 20	6 – 17·8	5·8 – 17·3
Other soils	.. ..	up to 2·5	0·7 – 8	0·6 – 4·6

These correspond to weights per acre of about 50 to 160 lb. on grassland and from 6 to 64 lb. on other soils. In discussing the relative consump-

[1] To convert to square yards deduct one fifth.

tion of bacteria by nematodes and protozoa Nielsen estimates that protozoa would be responsible for about 90 per cent and nematodes for 10 per cent.[1]

In view of the considerable quantities of potential food still remaining in the soil it may seem strange that the numbers of free living nematodes are not larger. The explanation probably is that much of the soil is inaccessible to them. They are not burrowing animals, but can only live in the pore space of the soil, and half or more of this may consist of pores of equivalent diameter of less than 20 μ, just too small for them to get into, (their diameters range from 20 to 50 μ) and therefore a safe refuge for bacteria, actinomycetes, small algae and other organisms that can enter. Moreover these smallest pores are filled with water, leaving only part of the total soil water for the larger pore spaces to which the nematodes must perforce confine themselves.

PARASITIC NEMATODES

These are far more important from the practical point of view than the free living forms because of the serious losses they can inflict. Their characteristics are that they spend their active life in the plant but some have a very resistant stage, the cysts, which can remain inert in the soil for several years; the eggs they contain are stimulated into activity by secretions from the roots of the host plant, as happened for certain parasitic fungi. In general each species has its own particular host and will enter no other, except perhaps a closely allied plant.

The first to attract serious attention were those which attack sugar beet. As is well known this crop was developed on the Continent by orders of Napoleon as one of his measures to counteract the blockade enforced by Britain during our wars with him. It proved so successful that it became one of the foundations on which German agriculture rested, and its cultivation continually extended reaching a high peak about the seventies of the last century. Then for no obvious reason the yields began to fall. There was no question of soil exhaustion; fertilizers were no cure, and farmers began to talk of the soil being tired of root crops (*Rübenmüdigkeit*). After much investigation J. Kühn in 1881 traced the trouble to eelworms which had already been found by

[1] C. Overgaard Nielsen's classical studies of the free living nematodes are published in English in *Natura Jutlandica* 1949, Vol. 2 which can be seen in some of the larger scientific libraries.

Schacht more than twenty years earlier, and which are now known as *Heterodera schachtii*. The eggs are contained in cysts in the soil; under the influence of excretions from the roots of the sugar beet they hatch out, and the young larvae, which are about 0·5 mm. long, make their way to the roots. Those that get there force their way into the cortex and stay there; they lie in the direction of the root, their heads pointing towards its upper end. There they feed. After about 14 days the males have become adult and are by now about 1·5 mm. in length but with some variation: they are very wormlike in appearance; at this stage they leave the root for the soil and proceed to find the females. These have swollen and become lemon shaped; they are shorter than the males (about 0·75 mm.) but much broader. In swelling they break apart the cells of the root cortex and while head and neck remain embedded therein their bodies protrude into the soil and are fertilized by the males. The eggs may drop into the soil, hatch out, producing larvae which then infect more plants, or the eggs may remain in the sac which toughens to form a resistant cover and becomes a cyst: here they may remain viable for some years. (Plate VIII, p. 115).

Larvae that do not at once find a root can survive in the soil for some months but cannot develop.

This eelworm occurs on the Barr.field mangold plots at Rothamsted on which mangolds have been grown every year since 1876. Periodical estimates of its numbers are made; they are highest on the plots that yield most heavily and progressively lower as the yields fall off. Over the period of observation there is no evidence that the population is becoming denser, nor do the plants suffer visibly.

The only method of dealing with the eelworm when it becomes a serious pest is to refrain from growing sugar beet or mangolds for some years. The interval between successive sowings may be four or five years in ordinary farm practice where arable crops and leys alternate; in those circumstances the eelworm population does not usually build up to a harmful level.

Of all this family the potato-root eelworm, *Heterodera rostochiensis*, is perhaps the most important in Britain because of the serious losses it causes, estimated about 1950 to be some £2 million annually for potatoes alone. But it also attacks tomatoes, and causes great loss to commercial growers.

Its home is in Peru; how it got here is a complete mystery. It was first found in 1913 simultaneously in Scotland and in Germany; four

years later it appeared in Yorkshire and in 1924 in Lincolnshire; it
now occurs in most of the potato areas of Great Britain and Ireland
and over most of Western Europe except Norway. So far it appears to
be limited to temperate climates.

The eggs are contained in cysts in the soil. The cysts are about
0·5 to 1 mm. in diameter and may hold some 600 eggs: they may remain alive but apparently completely inert for 10 years. Each year a
few may hatch out, but the great outburst is caused by excretions
from the roots of potato plants; these appear to be most copious at the
time of most rapid growth. The young larvae, about 0·5 mm. in length,
emerge from the cyst and enter the root, forcing their heads in by
means of the stylet which they can push out from their mouths. Here
they cause the cortical cells contiguous with the stele to become very
large, so impeding the circulation of the sap. Their subsequent be-
haviour is similar to that of the sugar beet nematodes already described.
The females begin to produce eggs which they keep within themselves
so that they swell considerably, becoming almost spherical. The head
still remains in the root cortex, but the swollen body is forced outside.
The males now leave the root and fertilise the females; then, their life's
work being done, they die. The body of the female, at first white,
becomes brown and the skin becomes very tough. After a time she dies,
leaving her bag of living eggs which drops off the potato at lifting time.
This is the cyst. The space of time between the entry of the root by the
female and the emergence of the swollen body is about five to eight
weeks.

The potato plant can tolerate a certain degree of infestation without
visible suffering: Peters[1] states that a healthy well-fed plant can carry a
population of many thousands of eelworms and show no signs of
damage; it simply responds by sending out new lateral roots. But
more usually the plant makes little growth, the leaves quickly die and
the tubers are few and small. On land where potatoes have to be grown
frequently a large population of cysts builds up in the soil and the
attacks become very serious, resulting in heavy loss.

The fact that emergence of the larva from the cysts is greatly
stimulated by an excretion from the potato root has encouraged the
hope that the active substance might be identified and prepared on
the large scale, then put on to the field to bring the larvae out of their

[1] B. G. Peters, *Rothamsted Annual Rpt. 1950*, 147–156. A very interesting description
of this eelworm from which much of the above account is taken.

eggs. As they are incapable of independent existence they would ultimately die (though they might take some months about it) if sufficient care had been taken to clear all potato plants off the land. The idea is very attractive and undoubtedly sound, and the Nematology Department at Rothamsted and Prof. Sir A. Todd at Cambridge have spent much time in trying to put it into practice. The excretion appears to be an unsaturated hydroxyacid of low molecular weight, but it is so unstable, and its quantity is so minute, that investigation is extremely difficult. Meanwhile the only remedy is to alter the rotation making a longer interval between one potato crop and the next, but this advice is particularly unwelcome in districts like the Holland division of Lincolnshire where potatoes contribute largely to the local prosperity and no less than one third of the arable land is occupied by them.

The typical potato soils, light, well aerated loams or fen soils, are well adapted to the eelworm as oxygen is necessary for the hatching of the cysts. On the heavier clay soils of Hertfordshire the cysts are not only fewer but smaller than in the regular potato districts.

So far no method of killing the cysts in the field has proved practicable.

Different species of the potato genus (*Solanum*) show different degrees of resistance to eelworms. This is important in that it opens up the possibility of breeding resistant varieties, but it is particularly interesting because of the possible explanations. Nothing definite has been established, but there are indications that the determining factor may be chemical: the resistant varieties may contain or produce some substance that prevents the larva developing when it has entered, or even kills it; or alternatively, the plant may fail to supply the larva with some substance essential to its continued development. No early solution of the problems of resistance can be expected.

The root-knot nematode which used to be called *Heterodera marioni* is now known to include several distinct species of the genus *Meloidogyne*. (Plate VIII, p. 115). It is a warm climate animal and does much damage in the tropics and subtropics. In England it flourishes in the soils of commercial nurseries and causes serious losses of tomatoes and cucumbers. It is dealt with by blowing steam through the soil before planting out. Travellers from Liverpool St. to Cambridge in passing through the great glasshouse region about Waltham Cross and Cheshunt may at times see a mobile steam generator at work outside the glasshouse; it is sterilising the soil in readiness for the young plants. Various chemicals are also effective.

The stem and leaf eelworms, *Ditylenchus*, *Aphelenchoides* and others, include a variety of different races attacking a number of plants, each usually having one chief host but some can do a certain amount of crossing over; the plants include oats, onions, potatoes, beans, red clover, lucerne, strawberries, bulbous iris and others: indeed before long it may perhaps be difficult to find a plant without an eelworm of its own. The black currant eelworm can in wet weather sometimes be found wandering along the branches: it can climb upwards against a downward trickle of water. Bulbs are sometimes infected, and the eelworms can persist throughout their period of storage and break into activity when the bulbs are planted and start growth. Fortunately they can be killed by heat: their death point is lower than that of the bulb, and by choosing the correct temperature and time of immersion in hot water the eelworms are killed without injury to the bulbs.

Some of the free living nematodes instead of feeding on soil micro-organisms feed on the plant juices and may cause considerable losses. They live outside the plant and are commonly called root browsers; they are provided with a stylet or mouth spear with which they penetrate the cortex of roots or underground stems and so are able to get at the plant juices.

Like all other soil organisms the parasitic nematodes have their predators: some of the larger nematodes can feed on them, and certain fungi can enmesh them in their mycelium and then proceed to parasitise them. Attempts have from time to time been made to modify the soil conditions so as to favour the predators but without success.

Mysterious bodies looking very much like eelworm cysts but much smaller have also been found in the soil. They are called microcysts and, being filled with cytoplasm, they should produce something living, but no one has yet succeeded in inducing them to do so.

The Enchytraeids

The soil enchytraeids are minute worms (mostly about 5 to 10 mm. long and 0·25 to 0·75 mm. diameter), weighing only about 0·01 mg. each when newly hatched and growing to 0·1 mg. or more at a later stage,[1] but they are very susceptible to dryness and few live to attain their maximum growth. C. Overgaard Nielsen in Jutland[2] found them, like

[1] These figures correspond to nearly 3 million newly hatched forms and 300,000 older forms per ounce. There are 28,350 mg. to the ounce.

[2] *Studies on Enchytraeidae*, Naturhistorisk Museum, Aarhus, 1955.

the Collembola and the mites, most abundant on acid heath soils (raw humus): the numbers were up to 200,000 per square metre;[1] on pasture soils he found about 50,000 to 100,000, and on arable soils only about 2,000 to 10,000 per square metre, the higher numbers where the soil contained more organic matter.

On the heath and pasture soils they remain very close to the surface, resembling the nematodes in this respect; on arable soils they go lower down. Their numbers were at a minimum in early summer: a little later many young larvae suddenly appeared suggesting they had hatched out from cocoons in the soil. From early autumn their numbers began to fall. They were not randomly distributed in the soil but tended to congregate in certain patches: Nielsen picturesquely likens this to the distribution of the human population in Denmark. The older forms, however, were more randomly distributed. They were much less numerous than the nematodes, being present only in thousands per square metre while the nematodes were counted in millions, but the biomass was not very different. Some of his results are given in Table 8.

TABLE 8

Numbers and weights of enchytraeids, nematodes, and micro-arthropods in certain Jutland soils. (C. Overgaard Nielsen). All per sq. metre.

	ENCHYTRAEIDS		NEMATODES		MICROARTHROPODS	
	THOUSANDS	GRAMS	MILLIONS	GRAMS	THOUSANDS	GRAMS
Site A ..	11 – 46	1 – 3	10	13·5	48	2
Site B ..	50	7		2	300	4 – 5
Calluna raw humus ..	40 – 200	5 – 25	1 – 3	1 – 3		
Conifer plantations	24 – 126	3 – 25	1 – 2	4 – 5		11

THE LARGE ANIMALS: EARTHWORMS

Earthworms are the largest and heaviest part of the animal population of the soil but although they have been known from time immemorial little interest was taken in them till comparatively recently. "Gardeners and farmers express their detestation of worms;" wrote Gilbert White in 1777, "the former because they render their

[1] 4050 sq. m. = 1 acre.

walks unsightly, and make them much work; and the latter because, as they think, worms eat their green corn. But these men would find that the earth without worms would soon become cold, hard-bound, and void of fermentation; and consequently steril." He sets out in this short letter (No. XXXV)—which is packed with sound observation—all that was for many years known of the effects of earthworms on the soil. He goes on: "Worms seem to be the great promoters of vegetation, which would proceed but lamely without them, by boring, perforating, and loosening the soil, and rendering it pervious to rains and the fibres of plants, by drawing straws and stalks of leaves and twigs into it; and most of all, by throwing up such infinite numbers of lumps of earth called worm-casts, which, being their excrement, is a fine manure for grain and grass."

It was not Gilbert White, however, but Charles Darwin who compelled the attention of scientists and gardeners to the important part that earthworms play in the soil. His book *The formation of vegetable mould through the action of worms with observations of their habits* published in 1881 is a model of patient observation and yet there is nothing that could not have been done by any interested person in his garden through the summer—if he had Darwin's keen perception and clarity of thought. His chief observations fall into two groups. Worms were kept in pots, and leaves of various plants were laid on the surface to see which would be chosen for food. Those of carrot, celery, wild cherry and cabbage were readily taken but aromatic leaves were not. Leaves were also used for lining the mouth of the burrows and for this purpose a wider choice was exercised including pine needles and even pieces of paper. Darwin also observed that the leaf or paper was seized by the pointed end rather than by the side, which struck him as remarkable in view of the fact that worms cannot see: he thought it indicated some rudimentary intelligence. Obviously this is the easiest way of getting the leaf into the burrow, but he noted that ants would often spend much time in trying vainly to drag along an object transversely, when it could much more readily have been moved longitudinally. Not only is the mouth of the burrow lined with vegetable material but it is also plugged with soil; a little heap of material might be erected over it.

Darwin's chief interest, however, was in the worm casts. He collected these on measured areas of land and found that in the course of a year they amounted to about 10 tons per acre. This provided quite a respectable covering of the soil and he records numerous instances where dressings of ashes or other material put on the soil a number of years previously were now buried several inches deep. The covering accumulated at the rate of about one-fifth of an inch a year and he discusses its widespread effects in producing the level sweeps of natural grassland in the chalk regions, the covering up of Roman remains, and other results. The estimate is probably too high: at Stonehenge the present rate of covering appears to be 6 or 8 inches per century.[1] In any case the burying is only in part due to earthworms. The chalk and the cinders which Darwin had spread on the surface in 1842 were buried to a depth of about 7 inches by 1871, but they were no lower in 1942 when Sir Arthur Keith looked for them.[2] Darwin's conclusion of chief interest for our present purpose was that "all the vegetable mould" (i.e. surface soil) "over the whole country has passed many times through, and will again pass many times through, the intestinal canal of worms". This still stands.

Much later work has been done, the most important in recent years in Great Britain is that by A. C. Evans and W. J. Mc. L. Guild at Rothamsted. Some 220 species of earthworms are known in Europe of which about 25 are found in Great Britain, 10 fairly commonly. Three of these are usually found in dung hills, and a few in mud banks.[3] Some half dozen species occur in the soil; two are of special importance because they alone normally make castings: *Allolobophora longa* and the large *A. nocturna*. All the British soil species are generally similar in structure and most of them are hard to tell apart. The body is cylindrical without head or neck but there may be a hundred or more segments, each with minute bristles which permit of movement backwards or forwards. A small mouth with prehensile lip but no teeth or jaws opens into the pharynx, beyond is the oesophagus in the lower part of which are three pairs of large glands which secrete quantities of

[1] R. J. C. Atkinson, Stonehenge (1956) Hamilton, London, (p. 55).
[2] *Nature*, (1942), 149, 716.
[3] Some of these mud worms are extremely interesting. One, *Eophila oculata*, common in Britain (though not in soil) apparently needs very little oxygen. It has been dredged from the bottom of a lake at a depth of 100 ft. and, still more remarkably, found in a ditch at Verulam which had been sealed over by floors of successive buildings during the first four centuries A.D. with no apparent way in or out. It looked as if the worms had been trapped there in Roman times and continued as a living community to our own day. They had plenty of organic matter as food. (R. M. Dobson & J. E. Satchell, *Nature*, (1956), 177, 796-7.)

calcium carbonate; then comes the crop where the food is stored pending transfer to the gizzard where it is triturated with the aid of particles of sand or grit, then it passes to the intestine where the nutrient substances are absorbed, finally it reaches the vent where all unassimilated material is excreted. There is no respiratory system: breathing is through the skin. There is, however, a circulatory system, also a nervous system and a highly developed sense of touch, but no organs of sight or of hearing.

Worms are far more numerous on grassland—lawns, meadows, etc.—than on arable land or in garden beds, partly because there is more food for them, partly also because they are not disturbed by cultivation operations. They are rare on dry sandy or gravelly soils and heaths, and also on badly aerated soils as they need oxygen, but they abound on chalk soils. On the Rothamsted grass plots they tolerate a certain degree of acidity but the limit seems to be about pH 4·5; on the strongly acid plots of pH 4 there are practically none. Accurate estimates of their numbers are not yet possible because directly the experimenter starts taking his samples of soil they take fright and burrow more deeply. An electrical method is being tried. They cannot hear, but they are very sensitive to vibrations: Darwin had noticed that worms emerging from their burrows in the pots in his study took no notice of sounds from the piano when they were alongside it, but retreated at once into their burrows when the pots were placed on the piano and a note struck—it was immaterial whether C in the bass clef or G above the line in the treble clef. At other times, however, when the soil is disturbed by forking or by a sampling tool they will quit their burrows and escape along the surface: there is an old country tradition that they do this when pursued by moles. The earlier Rothamsted estimates were up to 3 million per acre on old grasslands and about 1 million per acre on old arable land well manured (Broadbalk dunged plot); but fewer on poorer arable land. A. J. Low at Jealott's Hill obtained similar figures: 2·5 to 3 million on old grass land and 30,000 to 40,000 on old arable land.[1] Other workers prefer not to give a single figure but limits within which the odds are 20 to 1 that the true value lies—an odds of 20 to 1 in favour is the Rothamsted standard of acceptability in field experimental work. Evans' limits are given in Table 9.

[1] *Jl. Soil Sci.* (1955), 6, 179–199.

TABLE 9

The effects of previous agricultural history on earthworm populations of several fields; weight in lb. per acre, populations in 1000s per acre.

	WEIGHT TOTAL POPULATION	A. nocturna	A. caliginosa	Eisenia rosea	L. terrestris	A. chlorotica
Permanent pastures	646–456	459–285	228–141	141–60	33–14	58–35	17–4
Leys ..	643–443	323–219	61–34	92–70	28–14	97–68	
First year arable	746–362	330–230	198–109	67–38	18–5	53–25	20–3
Fifth year arable	125–61	129–69	31–13	46–2	29–7	4 or less	32–10

Hansen, the Danish investigator, has pointed out that the weight of earthworms on pasture land may be greater than the weight of live stock grazing upon it.

When grassland is ploughed or dug up the numbers of earthworms fall: slowly in the first year, more rapidly afterwards, to the new low level, the two large species especially suffering as shown in Table 9. The causes include lack of food, and easier access of enemies. When the land is put under grass again the population goes up. The numbers vary with the seasons: they are low during winter, they rise in spring, fall in summer, and rise again in autumn. (Fig. 7, p. 112). Except in summer the fluctuations in numbers follow the changes in temperature. The low numbers in summer may result from lack of moisture: worms were originally aquatic animals that invaded the dry land and they perish rapidly if deprived of water. The two important cast-makers, (*A. nocturna* and *A. longa*) go into a resting stage during the summer, curling themselves up in a cavity at the bottom of the burrow: this is called the diapause; two or three other species can also do this but need not. *L. terrestris* does not. Darwin described them as nocturnal in their habits but this is true only of three of the species, *A. nocturna*,[1] *A. longa* and *L. terrestris*.

Worms are hermaphrodite, each having male and female organs. Nevertheless they pair, the head of one towards the tail of the other.

[1] This, and *Lumbricus terrestris* (which is 5 inches or more in length) have value as fishing bait and enquiries are reported to have been made at Aberdeen for an export of ten million a year to the United States. It was stated that one man could collect a thousand a night on golf courses, parks, or open spaces (*The Times*, Dec. 29, 1955).

Spermatozoa are exchanged: they do not, however, immediately fertilise the ova but are stored for a time in a special sac. *L. terrestris* pairs at night on the surface; Gilbert White observed this and in the uncompromising language of the 18th Century declared that they were "much addicted to venery". Some other species pair in the burrows. The eggs are laid in oval shaped cocoons, produced by the swollen

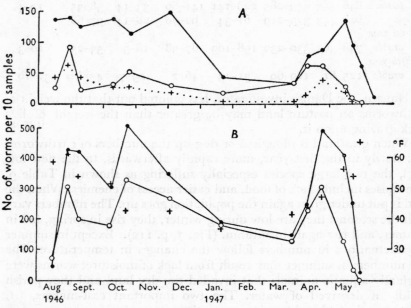

Seasonal trends in total populations. A: ●—●, *L. terrestris*; ○—○, *E. rosea* +...+ , *A. chlorotica*. B: ◒—●, *A. nocturna*; ○—○, *A. caliginosa*; +, temperature.

FIG. 7

Changes in the numbers of earthworms on Rothamsted grassland as the seasons change. (A. C. Evans & W. J. Mc. L. Guild, Ann. App. Biol. 1948, 35, 488).

girdle—the clitellum—of sexually mature worms: the cocoons also contain the fluid albumen that is to nourish the worm as it develops.

There is no special breeding season, and cocoons may be found at any time when the temperature is not too low, but mostly from March to November except during the period of diapause. *L. terrestris*, *A. longa* and *A. nocturna* have a much lower reproduction rate than some of the others. The time necessary for incubation in the cocoon varies a good deal and may be from 4 to 20 weeks; when the worm finally emerges it is fully developed with all its segments complete. But six to eighteen months have to pass before the worm is sexually mature. How long it can live after that is not known; in captivity, fed and watered and shielded from its enemies it has lived more than ten years. But the hazards of life in the field or garden are so great that no such longevity is at all likely: what with moles, birds, carnivorous insects and parasites life must be very difficult. The larger worms can burrow deeply to escape danger: *L. terrestris* can go down 4 or 5 feet; they can also spare a few segments if caught by a mole or a bird before their full length is safely in the burrow.[1]

Little is known of the food requirements of the different species of earthworm. Decaying vegetable matter is the chief food in natural conditions. Darwin observed that it was coated with a kind of alkaline saliva before being ingested—a unique instance of pre-digestion, he called it; as there are no teeth or grinding surfaces particles of grit are also ingested and the whole mass worked up in the oesophagus. He found that meat, fat and even dead earthworms were also eaten. In the Rothamsted experiments bullock, horse, and sheep droppings were all taken readily and induced greater cocoon production than cut grass: this is an important factor in the high population on grassland. Evans found five times as many earthworms on patches where dung had been dropped by the grazing animals some months previously as on patches where there had been no dung.

The burrows are made partly by forcing a passage through the soil, partly by eating a way through. The cast-makers eject the swallowed earth outside the burrow, the others eject it into some cavity in the soil. Any substance suitable for food would presumably be digested during the process. The mouth is so small that no particles greater than about 2 mm. (1/12th inch) can be ingested; the result is a considerable sorting out of the soil fractions by the earthworms.

Some estimate can be formed of the amount of earth passing through the bodies of the cast-makers; this is done by daily collecting and weighing the casts on a measured area. Darwin's mean was 10 tons per acre per annum with a range of 7½ to 18 tons; Evans' mean was about 11 tons per acre and the range 4·6 to 16·8 tons per acre. The top 9 inches of an old pasture soil may weigh about 1,000 tons per acre so

[1] See A. C. Evans, *Proc. Zoolog. Soc. London*, 1948, 118, 256.

that 11 tons per acre corresponds to about one tenth of an inch. There is however no means of estimating how much soil passes through the bodies of earthworms that do not make casts: the total is certainly greater than the figures here given. As a general guide Evans suggests that 4 to 36 tons per acre per annum pass through earthworms of which 1 to 25 tons are ejected as casts.

It is on grassland that earthworms exert their most important effects. Their burrows are channels down which water can drain away and through which air can penetrate and the thicker roots of the herbage can grow. The casts of the two casting species contain only fine mineral matter and hence where they predominate the surface soil tends to be enriched in silt and clay. But the intimately commingled organic matter gives the casts a better structure than the surrounding soil besides making them richer in plant food. These effects accumulate on old pastures long undisturbed by ploughing: Evans found that the soil of a very ancient pasture at Rothamsted had a greater pore space, fewer stones and less coarse sand in the top 3—4 inches than lower down, while on an old arable soil that had been in grass only seven years the differences were much less marked. (Table 10; see also p. 31).

TABLE 10

Effect of earthworms in sorting out soil and stones on grassland of different ages, Rothamsted.

	PARKLANDS	GREAT FIELD	PASTURES
Period in grass	OVER 300 YEARS	ABOUT 80 YEARS	7 YEARS
Weight of earthworms			
grams per plot ..	232	104	21
Worm casts, tons per acre	24·6	11·0	2·3
Weight of stones, tons per acre in successive layers			
0 – 2 inches	1	1	46
2 – 4 „	5	12	53
4 – 6 „	65	101	52
6 – 8 „	116	94	80
Weight of soil, tons per acre			
0 – 2 inches	170	198	265
2 – 4 „	243	237	294
4 – 6 „	285	274	279
6 – 8 „	312	370	323
Total ..	1,010	1,079	1,161

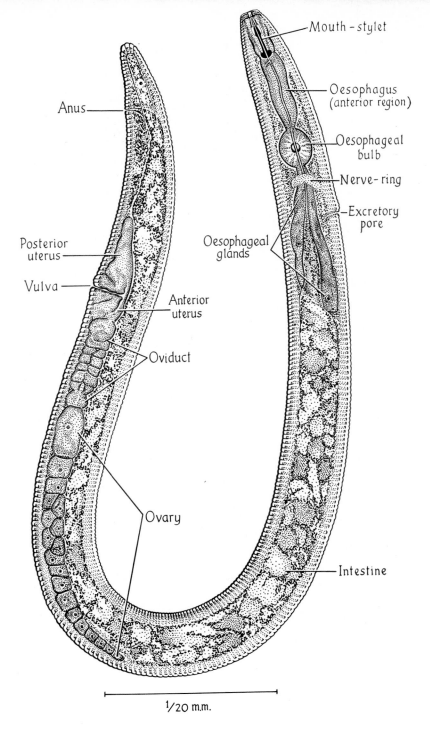

Anus

Posterior uterus

Vulva

Anterior uterus

Oviduct

Ovary

Mouth – stylet

Oesophagus (anterior region)

Oesophageal bulb

Nerve – ring

Excretory pore

Oesophageal glands

Intestine

1/20 m.m.

Plate VII. Meadow Nematode (*Pratylenchus pratensis*) × 900.
(Drawing by C. C. Doncaster)

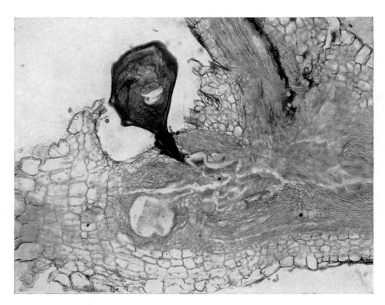

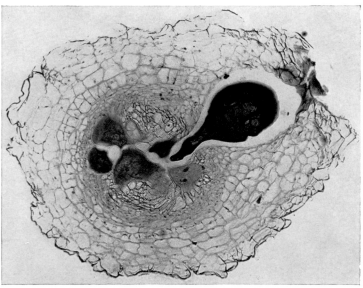

Plate VIII. Section of tomato root infested by nematodes: *a (top)*
Heterodera rostochiensis; b (bottom), Meloidogyne sp. × 65.
(*C. C. Doncaster*)

Parklands adjoins Great Field and there is no obvious reason except age why Parklands should be so much more heavily populated with earthworms than Great Field. It is known, however, that before 1870 Great Field had been for many years in arable cultivation and its surface as stony as that of Pastures. The accumulated layers of castings provided a four or five inch cover for the old stony surface of Great Field, and a somewhat deeper cover on Parklands, but the action seems to end there, and no appreciable amount of soil seems to have been brought up from lower down. The whole of the soil above this level has, however, been through the bodies of earthworms and its particles sorted out, the coarse ones being rejected. The comparative lightness of the top two inches and to a less extent of the second layer on Parklands and the two upper layers in Great Field is due partly to the greater pore space which weighs nothing, and partly to the accumulation of organic matter which is much lighter than the mineral matter.

The other important effect of earthworms, the removal of dead plant remains from the surface of the soil, is well illustrated on the Rothamsted Grass Plots. Those of only moderate acidity (pH greater than 4.5) carry an active earthworm population and are cleared of the dead leaves in autumn, consequently even delicate grasses and other plants can grow. But where the acidity is much greater earthworms cannot survive and the dead vegetation accumulates on the surface forming a mat through which only a few strongly growing kinds of grass can force a way and which ultimately becomes peat: as we shall see later this is how peat was formed. The difference between garden soil and peat is largely attributable to the action of earthworms.

These are long term effects and they show themselves chiefly on grassland. When we turn to arable land or garden beds regularly dug there is no clear evidence that earthworms play any important part in promoting soil fertility. They are much fewer and smaller than on grassland, the worm casters in particular are few and there is less for them to do. The mixing of soil with vegetable residues is done by the cultivations, the decomposition of the residues is readily effected by the micro-organisms, and the burrows are too few to play much part in aeration and drainage. No properly conducted experiment has shown unmistakably any short term advantage for the activities of earthworms on cultivated land. But the long term advantages remain, and the short term effects although not striking enough to be easily demonstrated, are continuously exercised for many months in the year: the aggregate effect may well be important. There is also the fact that earthworms are the predominant part of the macrofauna of the soil and it is impossible to foretell what the effect would be if they were wiped out.

Some American horticulturists have advocated the addition of worms to the soil so as to increase their effectiveness. Unless, however, adequate quantities of suitable food were added at the same time they could not survive; we have seen how drastically the population in grassland falls after the land is broken up. And if the suitable food is added and other conditions made favourable the native population will multiply to the appropriate new level.[1] There seems therefore little advantage in adding worms to the soil except that their dead bodies are a valuable fertilizer. There is however a better case for adding worms to a compost heap: they may hasten the decomposition of the coarse material.

Fears have sometimes been expressed that fertilizers injure the worms and should therefore not be used. There is no evidence of this even on the Broadbalk wheat plots, some of which have for more than a hundred years in succession received far heavier dressings of fertilizer than would be given in practice.

On special purpose lawns such as golf greens it is deemed necessary to reduce the worm population so as to avoid the formation of casts, and this is done either by watering in Mowrah meal or a solution of potassium permanganate. Unfortunately the treatment kills not only the two offenders, *A. longa* and *A. nocturna*, but the inoffensive ones as well and it deprives the lawns of the improved drainage and aeration provided by the burrows. Suitable methods of management have been worked out at the Bingley Research Station, Yorkshire and by Suttons of Reading.

One cannot think of earthworms apart from moles which seek them out and feed upon them voraciously, making long tunnels through the soil in their quest, at intervals throwing quantities of it up into little mounds. They undermine plants, and are an intolerable nuisance on lawns and garden beds, working especially along the edge between the bed and the grass. They breed about mid March: the young appear some six weeks later, generally about four to six in number. On semi-

[1] Application of birch litter to natural heath land enabled earthworms to colonise it (G. W. Dimbleby (1952) *J. Ecology*, 40, 331–341). Liming and phosphatic manuring increased the earthworm population in Lancashire pastures (Dobson, quoted by E. Crompton (1953) *J. Min. Agric.* 60, 301, 308).

wild grassland used for grazing they have some merit in that their tunnels aerate the soil, but on well cultivated land they serve no useful purpose. The older countrymen are often skilful at trapping them, but the younger generation has not acquired the craft, and the modern pest officer destroys the moles by placing in the tunnels worms killed with strychnine. Apart from the labour of catching the worms and poisoning them the method is probably very cruel, for strychnine poisoning is a painful death, and the mole is a mammal with a well-developed nervous system. The old trap had the advantage that death was instantaneous. A simpler and more humane method is very desirable: sticks of soft flexible gelatine cast in the shape of worms and impregnated with a quick poison might serve.

At a casual glance moles appear to have neither eyes nor ears but this is wrong. Their eyes are minute and almost hidden by the fur; the ears lack conches and are simply holes in the skin. Sometimes a movement of the earth will betray the presence of a mole and a rapid stroke with a fork may destroy it, but the slightest footfall suffices to put it on its guard.[1]

PLANT DESTROYERS

A certain number of animals living in the soil feed on growing plants and therefore come within the category of pests. Some of them contribute nothing to the building up of the soil as a habitat for plants and organisms and can be destroyed without compunction; others that feed also on dead plant remains or excreta of other animals could be defended as scavengers. The plant destroyers do not fit neatly into any of the Natural Orders but they are so important to farmers and gardeners that they must be included here.

The beetle family (Coleoptera) include both useful and harmful members. Some feed on dead plant material, rotting wood, fungi, manure, dead animal remains; some, the ground beetles, are carnivorous and feed on insects, worms and other small animals. Two at least must be condemned as pests because their larvae can be very destructive: the click beetle and the cockchafer; they give rise respectively to wireworms and white grubs.

Click Beetles: Wireworms
Three species of click beetle are common in Great Britain; all are

[1] See L. Harrison Matthews, *British Mammals*, (1954) in this series for further information.

small. One, *Agriotes lineatus*, is tawny coloured, has brown stripes running the length of its wings; another, *A. obscurus,* is earthy brown in colour; and the third and smallest, *A. sputator*, is black. They abound during the months May to July in grass fields, hedges and cornfields; *A. obscurus* also in gardens; they may often be seen towards evening resting on the tops of grasses. They are able to spring up when laid on their backs.

The eggs are laid close to a plant on which the larvae can feed when they hatch out. The young larvae are very small, only just visible to the naked eye; they feed voraciously and grow into yellow or yellowish brown segmented grubs with three pairs of legs at the head end. Their skin is extremely tough and they well deserve the name wireworm. They remain in the soil fairly near the surface from three to five years, moulting several times; then they go down more deeply, make a little cell, change into a pupa and remain till spring. As the soil warms up the beetles emerge, reach the surface, fly away; in due course they mate and more eggs are laid.

Wireworms are very numerous on old grassland and can readily be found by digging up some of the turf and pulling it to pieces. They are less numerous on old cultivated land because the hazards of life are so much greater there. They played a very important part in our national life during the two wars when millions of acres of grassland had to be ploughed up and sown with cereals. During the first war especially they effected great destruction; there was unfortunately no effective way of dealing with them. The older farmers had only two remedies: to roll the land so as to impede their movements and encourage the plant to put out fresh roots; and to grow a crop of mustard and plough it in, when it was said, the wireworms would feed upon it so voraciously that they would burst.

In the interval between the wars much research was done in the hope of finding a good soil insecticide that would destroy them, but without success. Methods were, however, worked out for estimating their numbers in an acre of land; for accurate work the principle of floatation (p. 93) was adopted, but for field work a sampling method was devised, the number of wireworms being counted in a specified number per acre of soil samples. The relative order of sensitiveness of the common crops was by this time known and a list was drawn up showing which could be grown, and which should be avoided, for given limits of population density. (Table 11).

TABLE 11
Tolerance of crops for wireworms.

NUMBER OF WIREWORMS PER ACRE[1]	SUITABLE CROPPING
Below 300,000	*All crops*
300,000 – 600,000	*Most crops except potatoes.*
600,000 – 1,000,000	*Peas, beans, flax (linseed); cereals and roots on heavy land only if in good condition.*
Above 1,000,000	*Peas, beans, linseed, grass seed. Dangerous for cereals and roots.*

A survey of the wireworm population made by the rapid field method showed that they were most numerous south of the Humber and in the eastern and midland counties, stretching out to the west in the south. Genera other than *Agriotes* were also found, especially in the north. Wireworms were more numerous in heavy than in light soils excepting fen soils and soils overlying the chalk where large numbers were found. Ample soil moisture during summer seemed to be important.

Considerable damage was done to cereal and other crops in the first two years after the grassland was broken up, especially sometimes in the second year. It was supposed that during the first year sufficient of the old grass roots survived to give an ample food supply, but as these decomposed the wireworms had to turn to the crops for food. Their numbers fall rapidly as the result of cultivation, but they rise fairly quickly when the cultivated land is laid down to grass: after three years a considerable population had accumulated at Rothamsted. Fortunately the predators are numerous and cultivation operations give most of them their chance. Many large birds take them: gulls, partridges, pheasants, plovers, poultry, rooks and starlings; some of the beetles are said to feed on them, and a small fly lays its eggs in them and the young maggots when they hatch feed on the body eventually killing their host. Mortality must be considerable, especially of the young larvae. Nevertheless the survivors can be very destructive: for each additional 100,000 per acre it was estimated that one hundredweight of cereal grain was lost.

The great development of technological organic chemistry during the war has provided a series of efficient soil insecticides, and so solved a problem that had long worried horticulturists and agriculturists.

[1] Only the larger ones (above 6 mm. length) were included because of the difficulty of finding the smaller ones. The flotation method gave much higher figures.

Gammexane, dieldrin and others destroy wireworms without injuring the crop; they are readily applied and can be used as seed dressings.

White Grubs

The Cockchafer deposits its eggs in the soil in clusters of 50 to 80; The larvae are dirty white in colour and remain in the soil for two or three years; when fully grown they are fat and about 1½ or 2 inches long. They are generally found in a curved position in the soil; they move but slowly. In lawns they feed on roots of grasses and plantains; in cultivated land they may do considerable damage to root crops and even to fruit bushes. When mature they burrow down into the soil to a depth of 1 or 2 feet, make a cell and pupate. In May or June the adults emerge but larvae can still be found intermittently throughout summer and autumn. Hedgehogs are said to dig for them on lawns in early summer. Gulls, rooks, plovers and starlings are very fond of them and pick them out from grassland and from soil upturned by cultivation. They appear to be more troublesome in France than in Britain; they can however be eliminated by modern insecticides: aldrin at the rate of 2—3 lb. per acre is said to be effective and it does not taint potatoes.

Other destructive larvae: the Leather Jackets

Leather jackets are the larvae of the familiar Daddy-long-legs; they are dirty brown in colour, they have no legs; their skins are tough and they are very destructive especially during the months May to August. When mature they pupate in the soil and the Daddy escapes in the autumn. The eggs are commonly laid in grassland: they are sometimes numerous in lawns and are among the many items of food which thrushes, blackbirds and other large birds seek to dig out in spring time. They are readily destroyed by the modern soil insecticides, DDT, gammexane and others.

Various flea beetles and their larvae do considerable damage at times.

OTHER SOIL ANIMALS

The Hymenoptera

Numerous species of hymenoptera occur in the soils: by far the most interesting are the ants, because of their highly developed social organisation. It has been said, though with some exaggeration, that they have passed through the same stages of social development as man, being first hunters, then pastoralists, and finally agriculturists. The most primitive are carnivorous; the more highly developed augment their flesh diet with honeydew, fruit juices, seeds and other plant materials; while other and presumably higher forms, cultivate fungi as food and are even said to keep their gardens weeded—these however do not occur in the British Isles but in tropical America.

The common British ants are of the hunting and pastoral type and are widely known because of their habit of "milking" the aphids that infest rose bushes and fruit trees: this is done by stroking them with their antennae and the exuded honey dew is then sucked into the ant's crop, taken to the nest and disgorged to feed the occupants there. Linnaeus called aphids the dairy cattle of the ants. If as not infrequently happens the aphids are carrying virus diseases the ants may assist in their spread and so do considerable damage.

Although originally all the individuals of the same group must have been alike they have now become highly differentiated. The great majority are the workers, wingless sterile females, some of which stay in the nest to tend the larvae while others go foraging. There are smaller numbers of fertile males and females: mating is in the air during a nuptial flight; shortly afterwards the males die and the females return to earth either to the old colony or to found a new one. A common yellow species lives entirely underground and makes extensive foraging galleries. Part of the structure may be uncovered when one moves a large stone in a garden or grass field. Some of the larvae are frequently seen on the surface of the soil but the workers hasten to drag them underground.

Ants are sometimes a nuisance to farmers by making anthills in meadows and so adding to the difficulty of haymaking, and at times they may be objectionable in a garden in sucking the juices out of stone fruit within their reach. Materials are on the market for destroying them. But a case can be made for their defence. They destroy considerable numbers of a wide variety of farm and garden pests. Indeed in Germany the Wood Ant, which lives in woods and makes its hill of pine needles and other litter, is protected and propagated because of its value as a predator on pests. These hills can be seen in our own woods: they may be as much as 3 ft. high and 6 ft. in diameter.

The different species of ants vary considerably in their actions and their foods—which are more varied than is usually supposed—and it is difficult to assess their relation to human economy.[1]

Centipedes and Millepedes. (Fig. 8, p. 124)

These two groups of animals must be considered together: they are not related but are often found in the same plot of land, and as one is frequently harmful and the other usually beneficial it is important to be able to distinguish them so as to ensure that the wrong animal does not get killed. Both are nocturnal and hide under leaves or stones in the soil during the day time.

Centipedes are carnivorous and feed on small insects and other animals, killing them first by poison which they inject through a pair of special organs. The name is a complete misnomer: most species have not got 100 legs nor anything like that number. Their bodies are long, flattened and segmented, each segment has one pair of legs. Six families are represented in Britain. In one of the two commonest, the glistening red-brown *Lithobiidae*, there are 14 segments; when disturbed these animals can move very rapidly. In the other common family, the *Geophilidae*, the bodies are longer and pale yellow and the movements sinuous; some of this group have at times taken to strawberries and become rather a pest.

Millepedes also do not deserve their name, but they usually have more legs than centipedes as each segment has two pairs; in spite of that, however, they are more sluggish. In the commonest species the legs are shorter than those of the centipedes; in some the bodies are cylindrical in others flattened. Unlike the centipedes they are mainly vegetarian though some occasionally feed on dead snails or other animals. The Spotted Snake millepede, formerly called *Iulus pulchellus*, now *Blaniulus guttulatus*, is very destructive, attacking roots, bulbs and tubers; it is about half an inch long, cylindrical, pale yellow in colour, with a double row of crimson or purple spots. Another species, *Tachypodoiulus niger*, is black and is also very destructive, especially of delicate tissues or internal tissues exposed by some other animal that has bitten away the protective cortex: their own mouth parts are not very powerful. One of the flattened forms, *Polydesmus angustus*, is common where rotting vegetation is abundant; it is about an inch long,

---

[1] I am indebted to Mr. M. V. Brian of the Nature Conservancy for information on this section. Those wishing to learn more should read "*Of Ants & Men,*" by C. P. Hoskins (1945) Allen & Unwin.

dull red brown and fairly active. The old method of dealing with millepedes was to trap them in pieces of mangold, which are then destroyed by burning, or by pieces of cabbage leaf soaked in Paris green which poisons them.

In spite of their bad character[1] they are extremely interesting animals. Like the Collembola they are very ancient: their fossil remains are found in the Devonian rocks and by Carboniferous times they appear to have been abundant. In all the millions of years that have elapsed since then they have changed but little in appearance. Yet once they were enterprising, for they were among the earliest animals to quit the water for dry land.

At times millepedes move in large numbers, even in daylight. S. G. Brade-Birks mentions an occasion when a train was stopped in Alsace because masses of them had been crushed on the lines, making them dangerously slippery.

Slugs

Slugs belong to the great group of molluscs but they differ from the majority in having no visible shell. They have, however, a small shell or granules representing one under the mantle on top of the front part of the body; this is usually regarded as a degenerate remnant. As a protection, however, the skin of some species is provided with glands from which when stimulated surprisingly large quantities of slime are excreted. The upper part of the body is rounded and in some species has a well marked keel; the lower part, called the foot, is flat. Two long horns or tentacles which can be drawn in serve as the organs of touch; two much shorter ones, rather behind the long ones, carry the eyes. The nervous system is something like that of earthworms, and concentrations of nerves scattered about the skin act as sensory organs. Slugs have a sense of smell, and are sensitive to light and to sound. On the right hand side under the mantle is a hole which serves for respiration and close to it are two others, one genital and one for excretion.

The digestive apparatus is similar in pattern to that of other animals. There is a mouth, furnished with lips and upper jaw; on the floor of the mouth is a ribbon of chitin called the tongue or radula containing many recurved teeth. The jaw and radula rasp off fragments of leaf or other material, and a pair of salivary glands pour saliva on to the mass.

[1] Like other bad characters they sometimes serve a useful purpose: they have acted beneficially in forest soils (L. G. Romell (1935) *Ecology*, 16, 67–71).

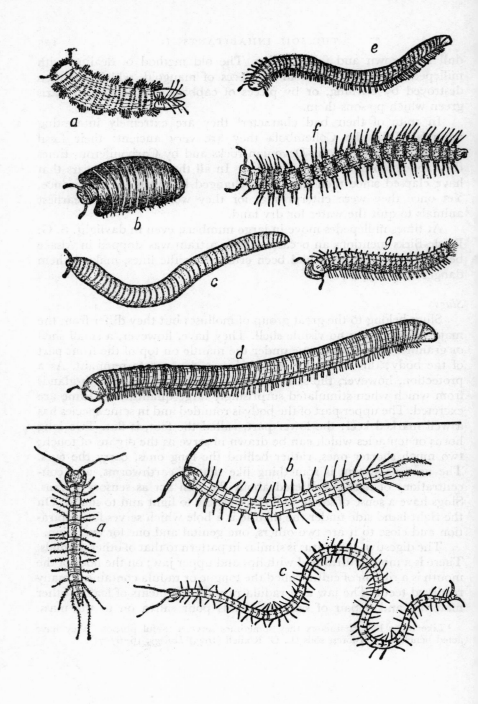

This then passes along the oesophagus into the crop to await digestion, then into the small stomach into which two large brown glands, the liver, secrete the digestive enzymes and promote the absorption of digested food for storage and the excretion of waste substances from the blood. From the stomach the residues pass along the intestine to the rectum and on to the anus where they are excreted. In most animals the digestive organs run in more or less straight line from the mouth at one end of the body to the anus at the other. But the slug is different. At its hind end the alimentary canal, instead of terminating, doubles back on itself and the anus is in the front part near to the respiratory and the genital pores.[1]

The reproductive system varies in the different species. Each individual usually possesses both male and female organs but they do not in general develop simultaneously, although this sometimes happens; occasional individuals are unisexual. Mating is however, the general rule. The eggs are small and white, they are laid in batches: one of the most prolific, the Grey Field Slug, has been observed to lay 425 eggs in four weeks.[2] Hatching out takes some three or four weeks. The young slugs fairly quickly become adult; they can survive the winter and one species, the big black one, is said to be capable of living 3 or 4 years.

They are nocturnal in their habits, and are normally active from about two hours after sunset to two hours before sunrise: the Garden Slug, however, oversteps these limits and will begin operations before sunset and go on after sunrise. In gardens the day is spent in hiding under stones, in rockeries, in loose brickwork, or by some species, down in the soil: in the field the soil is the usual refuge. Slugs are not usually killed by frost though it immobilises them, but as the temperature rises they become active and by the time it has reached about 40°F. all are fully active. They cannot, however, operate in dry conditions and many perish during a dry summer; they do not like heavy rain or high

[1] *Testacella*, a species that feeds on earthworms, is an exception.
[2] Between Nov. 25 & Dec. 22, (R. Carrick, (1938), *Trans. R. Soc. Edin.* 59 (3), 563–597.

FIG. 8
Opposite above: Millepede families. a. *Polyxenidae*. b. *Glomeridae*. c. *Polyzonidae*. d. *Iulidae*. e. *Blaniulidae*. f. *Polydesmidae*. g. *Craspedosomidae*.
Below: Centipede families. a. *Lithobiidae*. b. *Cryptopsidae*. c. *Geophilidae*.
Not to scale. Drawn by J. L. Cloudsley-Thompson.

winds, but do not seem much affected by ordinary fluctuations of temperature. The ideal conditions for them are warm still nights when the surface of the soil is nicely moist from recent rain or dew. Surprisingly large numbers can be caught by a person with sharp enough eyes and an electric torch; in the course of half an hour H. F. Barnes has in his own garden caught 570, then after an interval of half an hour another half hour's forage yielded 517.

They can take a great variety of foods so long as these are on or near the surface of the soil, and they do considerable damage to low lying leaves and crowns of plants like primulas: they prefer damaged or partly decomposing plant material. They seem particularly attracted to meat and bone meal, Quaker oats, and bran, so much so that a mixture of any of these—especially bran—with metaldehyde (Meta) is an effective death trap for them so long as it can be protected from heavy rain.

There are some 20 species in different parts of England: about a dozen are fairly common. The most comprehensive survey yet made is that by H. F. Barnes and J. Weil of Rothamsted in some of the gardens at Harpenden in the years 1940—43 in the course of which some 100,000 slugs were dealt with. Nine species were found, five were fairly numerous. The two commonest were the Garden Slug, *Arion hortensis*, and the Field or Grey Field Slug, *Agriolimax reticulatus;* in some gardens one preponderated, in others, the other.

The Garden Slug is up to about ¾ inch long, but almost double that length when moving; it is black or rusty black in colour; the foot is yellow or orange. It breeds all the year round, but probably most frequently in autumn and winter. Mating is on the surface of the soil but usually under some cover such as is provided by a dead leaf. Its numbers are at their highest in the period October to December and fall to a minimum in dry summer months, but it is very sensitive to weather conditions.

The Grey Field Slug, probably the best known of all the group, is small, pale whitish to brownish in colour, mottled or speckled. It exudes a white slime when touched. It is not confined to the surface for feeding but can climb up the plant. Although common in gardens its true habitat is probably grass or cropped land; sometimes it does great damage to young cereal crops. It is the hardiest and most resistant of all the slugs. It breeds throughout the year, mating freely in the open, and attains its maximum numbers in late summer or early autumn; its numbers fall considerably in the November period when the accompanying Garden Slug is most plentiful.

Another common slug is *Milax budapestensis*, about 2 inches long, very dark brown or grey black,[1] it has a well marked keel on its back and its sides are lighter in colour. It does not exude a slime but is very sticky. Like the Garden and Field Slugs it breeds throughout the year, and samples collected during any month contain individuals at all stages of growth. Unlike these two species it feeds below the surface and in particular attacks potato tubers in autumn: it is one of the few species of slug that can break through the potato skin. It can do a great amount of damage: Dr. Barnes found a potato patch on which all the tubers had been consumed, the skins only being left: the numbers of slugs found suggested that they must have eaten something like 30 to 40 times their weight of potato, which seems to imply that they are inefficient feeders. Other foods include puff balls, toadstools, bones, dog faeces, squashed slugs: Barnes found some of them feeding on slugs that had been killed by Meta—but the consequences were fatal. Mating takes place underground. Their numbers are highest in the September—November period. Like the Garden Slug they are very susceptible to weather conditions.

The closely related *Milax sowerbyi* is larger, usually about 3 inches in length, brownish, minutely mottled with orange or yellow and black. It is also an underground feeder and can penetrate the skin of the potato tuber though it prefers puff balls. When the Milax has made a way into the potatoes the Field Slug can follow and effect its full share of destruction.

Another species, *M. gagates* resembles *M. sowerbyi* in size, in being flattened from side to side, in humping when touched and in destructiveness. It differs in having translucent colouration and a head darker than the rest of the body, while that of *M. sowerbyi* is paler or orange coloured.

Arion subfuscus, the Dusky Slug, is large, about 3 inches or more in length, and brown. It is perhaps the least culpable of all the common slugs, for it seems to feed chiefly on fungi and on faeces of various animals though it has been observed to take the leaves of garden plants. It is most numerous and active during the summer, Like the Field Slug it mates in the open, chiefly in the period July to October, with a

[1]The colours given in this section are those recorded by Barnes at Harpenden. He points out, however, that they may vary in different localities and habitats.

peak in August. Babies are most abundant in December and January: by the following September the large adults are beginning to appear moth-eaten and it is doubtful if many survive a second winter. It is really a woodland slug, but Barnes regularly found it in Harpenden gardens in 1942 and 1943.

The only other slug that need be mentioned is the big black slug, *Arion ater*, common in coarse grass meadows and verges, but happily not numerous in gardens free from coarse grass. It is not born black: when newly hatched it is white to orange coloured, but by the time it is an inch long it is already black. When fully grown it is about six inches long. The younger ones predominate during the early months of the year, but by June and July nearly all specimens found are adult. In late autumn, however, young ones reappear and throughout winter form the bulk of the population. The breeding season is during autumn.

A. ater is not the largest species in the country: another, happily much less common, *Limax cinereoniger*, may grow to more than 15 inches in length. It is purely a woodland slug, feeding on fungi, and in spite of its terrifying appearance, is harmless.

It will be noted that the various species in one and the same garden do not reach their highest numbers at the same time. First comes *A. ater*, at a maximum in January[1]; then in early summer, June and July, *A. subfuscus*; then throughout the autumn come first the Grey Field slug, followed quickly by the two *Milax*; and lastly at the end of the year, the Garden Slug, *A. hortensis*. The breeding peaks are at about the same times, and from the New Year onwards the young slugs are growing up. A garden with a slug population is therefore not likely to be without them at any time of the year. The minimum period of activity (apart from frost) is midsummer when, except for *A. subfuscus* slugs are neither large nor numerous. Damage is most obvious in the spring because the only foods available are the young and tender shoots of plants. The consumption of plant substance, however, is greatest in autumn and early winter, but by that time there is much unwanted material in the form of decaying leaves so that the plants suffer less. The period of greatest activity is round about midnight: the Field Slug *Ag. reticulatus* rather before, *M. budapestensis* after and *A. subfuscus* in between; but the Garden Slug, *A. hortensis* keeps longer hours, as already stated.

[1]Being young and only straw coloured they may be overlooked in January; they do not become black and conspicuous till late summer or early autumn.

The different species vary in weight: 130 full grown Garden Slugs go to the ounce; the Field Slug and *Milax* are about 50 per cent heavier, while a Black Slug may weigh about ¾ oz. Barnes estimates that the total weight of slugs in a garden may be of the order of 180 to 300 lb. per acre.

If they all consumed food at the same rate as the *Milax* on the potato patch mentioned above the total would amount to some $1\frac{1}{2}$ to 5 tons per acre of plant material.[1] The higher figure seems improbable but the lower figure may not be far out. They are unquestionably a nuisance in a garden, but something can be said in their defence: they are great scavengers. Fortunately slugs have their enemies: toads, grass snakes and slow worms eat them, while their fatal passion for bran makes trapping with a bran-Meta mixture effective, provided it can be kept dry under some protecting roof.

Snails

Gardeners and countrymen distinguish between slugs and snails but actually there is no biological distinction; the only difference is that the snail's shell is big enough to house him while the slug's is not. During winter snails retreat to some sheltered spot, close up the entrance to the shell by a layer of mucus which speedily hardens, and remain safely sealed in till the warmth of spring stimulates them to activity. Slugs and snails are members of the same family; their internal structure is the same and their mode of reproduction. Both have the same large crop in which the ingested food is stored prior to passing into the small stomach for digestion, for both have the same habit of eating in haste and digesting at leisure in some quiet spot. Both have the gland in the forepart of the foot that excretes the slimy track along which movement is practicable without injury to the delicate foot. Both are hermaphrodite but mate for the exchange of spermatozoa and both are equally dependent on moisture. Snails are, however on the whole less destructive in field and in garden than slugs. Only two species can be regarded as pests, the Common Snail, *Helix aspersa,* and the Strawberry Snail, *Trichia striolata.*

The Common Snail has a fawn coloured shell, banded and marbled with dark brown; it is thick, nearly globe-shaped, but somewhat flattened, and is about 35 mm.[2] in height and in breadth. It lives

[1] The dry weight would be about one eighth of this.
[2] There are 25.4 mm. to the inch.

amid thick herbage, especially ivy and evergreen shrubs, under hedges and in sheltered parts of the garden. It has a fixed home to which it returns after each foraging expedition; Ellis records instances in some limestone districts where deep holes have been worn in the rock by countless generations of snails making their homes there for thousands of years.

It used to be eaten; numbers of shells have been found at various sites indicating that it was a regular article of diet in Romano-British times. The custom persisted almost to our own times in some parts of the country: Ellis quotes statements to the effect that it used to be sold in the Bristol markets as "wall-fish" and was much relished by the poor of Bristol, Swindon and other towns.

The more highly esteemed edible snail that figures in some of the Paris menus as *escargots de Bourgogne,* is the somewhat similar but larger *Helix pomatia,* the so-called Roman snail, often said to have been introduced into this country by the Romans, though as a matter of fact it was here before they came. It is about 45 mm. in height and in breadth, pale yellow in colour, marked with 3 to 5 pale bands. Numbers of shells have been found in the vicinities of Roman villas: the shells of the Common Snail, however, are often confused with them.

The other destructive snail, the Strawberry Snail, is typically small and squat, about 13 mm. wide but only 8—9 mm. high. Two forms are common: a flat but rather larger one in the east and south-east of England, and a smaller one possessing a rather higher spire in the west and north-west. The shells vary in colour; they may be white, dusky yellow, dark reddish-brown, or some intermediate shade. Like the Common Snail it is found in abundance among dense vegetation, ivy, nettles, hogweed, and damp waste places generally; it is apt to be troublesome in gardens and as its name implies, it may do considerable damage in strawberry beds.

Another species, the Garden snail, *Cepaea hortensis,* is larger, being about 15 mm. high and 18 mm. broad; it also occurs in two forms which are about equally common: both are yellow but one has five dark bands and the other has none. It is common in gardens, and it eats a variety of plants including young and tender vegetables such as lettuce, but prefers nettles, ragwort, hogweed; it cannot be regarded as a serious pest. It is gregarious and lives in colonies in moist sheltered places with abundant vegetation cover. Thrushes are very fond of it, and a little pile of empty shells may not infrequently be found near the stone on which they were beaten in the efforts to extract the snail. The task cannot be quite easy, for the shell is not a separate thing like a house but an actual part of the body; it grows with the body so that there is no need for moulting.

In view of the great number of species of slugs and snails in the country it is consoling to think that only a few do appreciable damage to cultivated plants: the Field Slug, the Garden Slug, the peculiarly flattened Bourguignat's Slug, and the three Milax species; the Common Snail and the Strawberry Snail. The situation might easily be worse.[1]

A curious feature of both slugs and snails is that their blood contains a copper compound, haemocyanin, which gives it a faintly bluish colour, instead of the red haemoglobin characteristic of the higher animals. It would be interesting to know how they get the copper, for plants contain only minute quantities and some soils are known to be very deficient in it.

There still remain two animals that should be included in this chapter, although they play no important part in the economy of the soil: earwigs and woodlice.

Earwigs are nocturnal in their habits: they feed on tender foliage, retarding the growth of ornamental plants in the early stages and spoiling the flowers later on by nibbling the petals, they are particularly attracted to dahlias. The female lays eggs in masses of 50 or more in a cavity in the soil during winter and early spring; she remains with them till they are hatched. The young larvae develop in the soil moulting four to six times: finally they emerge during the summer as adults. They then have wings which, however, they rarely seem to use. Their habit of seeking cracks and crevices enables gardeners to trap them in inverted pots stuffed with straw and mounted on canes.

Woodlice have little association with ordinary mineral soil, but in fen and marsh soils they are the chief agents in breaking down plant debris, replacing earthworms in this respect.[2] They are however of the same great family as crabs and shrimps but they took to life on the land. A point of special interest is that the females instead of depositing the eggs in the soil extrude them into a pouch on the underside of the thorax and keep them there till the young woodlice hatch out.

[1] The reader wishing for more information about slugs and snails can find it in an interesting volume by A. E. Ellis, *British Snails*, (1926), Oxford, Clarendon Press.

[2] O. Cooper, *Natural History of Wicken Fen*, Chap. XII p. 148 and A. E. Ellis, *Trans. Norfolk & Norwich Nat. Soc.*, Vol. 15, p. 291.

Woodlice live in decaying wood, dead leaves, and damp places, and they may do considerable damage to plants in hot houses, frames etc. Small infestations can be dealt with by trapping, and seriously infested glass houses can be cleared by fumigation.

The first thing that strikes one about the meiofauna of the soil is its amazing variety. The largest—the big earth worm Lumbricus, is of the order of one million times the size of the smallest, a mite. The smaller the animal the more numerous it is: Salt and his colleagues found that the numbers of arthropods in the soils they examined were in the following order compared with those of the largest:—

Smallest: size of mites and Protura   27
Larger: size of Collembola   10
Still larger: size of Soil Aphids (Symphila) and Thysanura   4
Largest: Up to millepedes and centipedes   1

The larger ones and the larger members of the different species tended to remain in the upper six inches of soil, and the smaller members of the species in the lower six inches.

The age structure of a given species tends to be pyramid-shaped. There are large numbers of the very young, and rapidly decreasing numbers as the age increases, indicating a high mortality risk.

All of the animal population, like the micro-organisms, live entirely on plant materials either directly or indirectly, their numbers and species in the soil are indeed largely determined by the vegetation. A change in the vegetation such as would result from a change in the management of the soil, the addition of lime or manures, or the introduction of a so-called soil improving tree on wild heath land, may set in train a sequence of changes in the soil population far greater than could have been anticipated: the increase in numbers of earthworms has been mentioned on p. 111, many consequential changes in the fauna follow.

The soil animals serve the useful purpose of helping to consume the mass of vegetable débris which, if it accumulated, would bury the soil and in time completely change its nature. Animals and micro-organisms thus achieve the same end: the two sets of operations go on simultaneously and with some interlocking. The products are, however, different. The starting point for both is vegetable matter with a carbon/nitrogen ratio of 20 or 30. Fungi and bacteria reduce this to relatively stable humus with a carbon/nitrogen ratio of about 10. Animals on the other hand build up tissues rich in nitrogen and easily decomposed; their total mass in the soil is a reservoir of nitrogenous matter continuously being broken down but continuously replenished, and constituting an appreciable proportion of the total soil nitrogen. The lives of many of the animals are short and an atom of nitrogen entering into the animal cycle may form part of several different combinations in the course of a year. There is no evidence that animals can make humus.

Little is known of the part played by individual groups excepting the earthworms. These are large enough to be picked out easily: pots of soil can be put up entirely free from earthworms and compared with others to which earthworms have been added, and their activities can thus be studied. Experiments on these lines have shown that the main effect of earthworms is to drag the dead plant residues into the soil, break them up into fragments and mingle them with the fine mineral particles that they are able to ingest. Investigations on similar lines with the smaller animals are impracticable because they can never be entirely removed from the soil; it is impossible to sterilise a soil without fundamentally altering its properties, consequently one cannot start with a soil completely devoid of animal life, and proceed to add the various species one by one to see what effects they produce. A certain amount of information has nevertheless been obtained.

The most numerous—and also the smallest—of the meiofauna, the mites and the Collembola, probably effect considerable comminution of the vegetable matter relative to their size. They have a high rate of metabolic activity judged by the oxygen they consume; but there is no clear evidence that they can decompose lignin or cellulose apart from what might be done by bacteria living in their gut; they obtain sustenance from these substances *via* the fungi on which they feed. Their excreta so far as investigations have yet gone consist largely of finely divided plant material from which they have extracted such substances as they could digest. Murphy points out that the aggregate amount of comminution they could effect in a forest or heath-land soil covered with litter could be considerable. Compared with soil fungi which can assimilate 30 to 50 per cent of the carbon they transform, these minute animals are extremely inefficient feeders, nor are the larger ones much better.

When however the transformations effected by the animals are regarded as a whole they are seen to be much more economical than appears at first sight. An animal retains only a small proportion of the material it ingests during its life time: some is oxidised to provide energy, the carbon being converted into carbon dioxide and the nitrogen excreted probably as uric acid. The greater part of the ingested material is, however, excreted without undergoing drastic change. But here is where the economy of the process comes in. The excreta are not wasted: they constitute the food of the coprophagous members of the soil population, which is virtually a cross section of the entire population including members of all the great groups: nematodes, beetles, and larvae of many kinds. These individually are probably no more efficient feeders than the original transformers, but their excreta in turn serve as food for other animals and for micro-organisms: so long as any energy-giving material remains some form of living organism proceeds to feed upon it. B. R. Laurence showed that cowpats harbour a rich population of small animals and he estimated that during a year a bullock leaves in its faeces enough food to support an insect population of at least one fifth of its own weight, to say nothing of the bacteria, fungi, protozoa, nematodes and other organisms that also participate.

Meanwhile the animals that have consumed these various materials retain them only temporarily: the hazards of life are great, they are attacked by parasites from within and predators from without; mortality especially in the early stages is very heavy. Overgaard Nielsen distinguishes three great cycles:—

I. *Organisms feeding on dead organic matter.* Bacteria, actino-mycetes, fungi, various minute animals, certain larvae, earthworms, millepedes and others.

II. *Organisms feeding on those in Cycle I.*
    On bacteria: certain protozoa and nematodes;
    on fungi: certain mites and Collembola;
    on actinomycetes: certain nematodes;
    on enchytraeids and earthworms: certain insects and vertebrates.

III. *Organisms feeding on those in Cycle II.*
    On protozoa: certain nematodes;
    on nematodes: certain nematodes and mites;

on mites: certain mites;

on insects: certain insects and vertebrates.

His diagrammatic representation cycle of nematodes is given in Fig. 9 (p. 136).

The animals do not appear to complete the degradation of the organic matter: much complex material remains in their excreta, the nitrogen compounds are apparently not broken down beyond the stage of uric acid. Bacteria, however, are able to go further; their end products are carbon dioxide, nitrates and the mineral salts already mentioned: these are quite unable to sustain animal life, and they represent the final stage of the chain of animal activity: the cold end points. Here again, however, the same economy appears: these substances are the essential nutrients which enable plant life to develop and to produce more plant products which in turn enable the animal population to thrive and multiply. So the Wheel of Life in the soil grinds on endlessly.

Unfortunately there are no good figures showing the total masses of micro-organisms and of meiofauna in British soils: the best available refer to a Swiss grassfield and are given in Table 12.

TABLE 12

Weight of soil population to a depth of 6 inches in a Swiss grass field, lb. per acre:

Microflora	18,000
(*including bacteria*	9,000)
Protozoa	337
Nematodes	44
Enchytraeids	13
Earthworms	3,560
Mites	
Collembola	
Protura	10
Diplura	
Other invertebrates	712
	22,676

(Stöckli, A. 1946, *Ber. Sch. Bot. Ges.* 53A).

In arriving at these results various assumptions have had to be made in

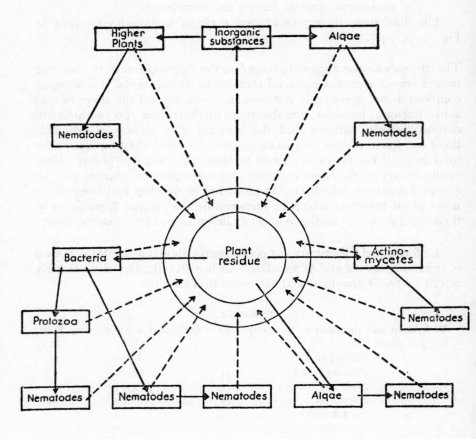

FIG. 9

The food cycle of free-living nematodes. The outer circle represents the soil organic matter; the inner circle represents the part due to plant residues and the surrounding ring is the contribution made by the dead bodies of the soil population. (C. Overgaard Nielsen).

————————	"enter into".
– – – – – – –	"turn into when dead".
— — — — —	several stages involved.

regard to average weights and the figures have little more than illustration value: they are however consistent with results obtained elsewhere. Lower figures would be expected for arable soils.

Although the part played by individual members of the soil population cannot be determined some estimate is possible of its over-all activity. Reference has already been made (p. 27) to the losses of organic matter from the plots in Broadbalk wheatfield, Rothamsted, one of which receives annual dressings of farmyard manure while another does not. E. W. Russell[1] has calculated the losses in terms of kilo-calories per acre per annum:—

	Plot receiving farmyard manure	Unmanured plot
	millions	*millions*
Added in manure and stubble	16	0·3
Remaining in soil	1	—0·7
Dissipated	15	1

The calories lost have mainly been consumed by the soil population. No recent figures are available showing the total numbers of all living forms on the two plots, but where determinations of particular groups have been made the numbers on the manured plot have been about three or four times those on the unmanured plot. Yet the energy dissipated has been fifteen times as great: evidently there is a large luxury consumption of energy where considerable quantities of organic matter are available for the soil population.

It is interesting to note that the 15 million calories lost from the soil and presumably consumed by the soil population would have sufficed for the needs of 12 persons, while the calories in the human food produced sufficed only for two. Much of our agricultural effort goes in sustaining the large and varied population of living things in the soil: we get only the by-products of their activity.

Soil Conditions & Plant Growth, 8th Edn. 1950, (p. 194).

CHAPTER 6

MAN'S CONTROL OF THE SOIL: I. PLANT
FOOD: ORGANIC MATTER

IT WAS early discovered that people who wanted to grow particular plants must look after them: if left to Nature they might easily come to grief. Before the dawn of history it was known that manuring and cultivation were essential for success and both practices were developed empirically throughout the ages, finally attaining in the 1860's and 1870's high standards of craftsmanship. Happily some of this still survives, and the green-fingered gardener is still with us, exciting our envy and admiration by the way in which he (or very often she) can induce even the most reluctant plant to flourish. Craftsmanship alone, however, has long proved insufficient to give the degree of control of the soil now necessary, and fortunately science has developed sufficiently to indicate ways in which greatly enhanced growth may be attained.

The question: "What makes plants grow?" has always interested thoughtful people. Much experience based on the marked manurial effects of decaying vegetable and animal matter in the soil was summed up in the ancient saying "Corruption is the mother of vegetation", and Chapter IV shows the basis of truth in this. The early scientists, wishing to be more precise, sought a "Principle of Vegetation". Van Helmont in the 17th Century thought he had found it in water—as the old Greek philosopher, Thales, had done long before him. Glauber, somewhat later than Van Helmont, thought it was saltpetre. Jethro Tull in the 18th Century considered that plants fed upon the fine particles of soil, while the chemists and botanists of the time regarded the juices of the earth as the proper food, and that plants took them up much as animals do, the roots being "but guts inverted". There was a large element of speculation in all this, as indeed was inevitable until chemistry and plant physiology became much more highly developed.

The agricultural chemists were the first in the field. They showed that plants were made up of carbon, hydrogen, oxygen, nitrogen, phosphorus, and a group then called the alkalis and alkaline earths: potassium, magnesium and calcium; other constituents have since been discovered. The first three were soon shown to come from the carbon dioxide of the air and the rain, nothing need or could be done to control them. But the others came from the soil or the manure. Two deductions were drawn: neither is correct, but both persist to this day in some form or other. The fact that a substance was found in the plant was taken as evidence that it was serving some essential purpose; and manurial formulae were drawn up on that basis. Since carbon dioxide and water are available in large amounts in the air and the rain there is no need to add them in the manure, but the mineral elements are needed. The other deduction was that, as one ton of farmyard manure contains only about 30 lb. in all of nitrogen, phosphorus and the potassium, calcium and magnesium group, it is more effective, as well as being more economical, to supply these in the forms of simple salts which are readily available in large quantities and at low cost.

These views were set out by the distinguished German chemist Justus von Liebig in the 1840's and controverted by Lawes and Gilbert at Rothamsted; the controversy died down but has several times been revived. Its immediate effect was to interest chemists in the simple salts containing the elements required by plants: nitrate of soda and sulphate of ammonia containing nitrogen; calcium phosphate containing phosphorus; and the alkali salts. The first two had long been used as fertilizers. Nitrate of soda came from Chile, where it occurs in enormous deposits though no satisfactory explanation of its origin has yet been given: no comparable deposits are known anywhere else. Sulphate of ammonia was a bye-product of the gas works which, from 1820's onwards, began to spring up in many parts of England: the gas contained a certain amount of ammonia which was not only useless but unpleasant, it was therefore removed by sulphuric acid. It was, however, calcium phosphate that brought about the revolutionary change and led to the establishment of a new industry: the manufacture of fertilizers.

The beginning was almost accidental: a chance conversation between John Bennet Lawes, the young squire of Rothamsted, and a neighbouring landowner, Lord Dacre, some time in the late 1830's. There was at that time much discussion about the manurial value of ground bones: on some soils they acted splendidly; on others, including Rothamsted and the neighbouring farms, they did not. Lawes was interested in chemistry and had in fact converted a barn into a laboratory, though he did not study farm problems, instead he was concerned with medicinal plants. Lord Dacre, however, asked him to try and find out the reason for the inefficacy of bones on their farms. Lawes was at once interested. He realised that plants can take up only dissolved substances: bones, he knew, were insoluble but they could be made soluble by treatment with sulphuric acid. He prepared a quantity of dissolved bone and found it acted well on his soil.

This discovery would not by itself have been very important: bones were scarce, and British dealers were searching high and low for supplies; indeed Liebig in an angry outburst declared that they were ransacking the battlefields of Europe. By good fortune it happened that deposits of calcium phosphate were at that time being discovered in Western Europe: it was useless as manure as then ground, but after treatment with sulphuric acid it proved effective, especially on the turnip crop, then very important in British farming. So in 1842 Lawes patented his process, in the following year he set up a factory at Deptford Creek and started the fertilizer industry, greatly to the benefit of British agriculture—and incidentally of himself.

For over seventy years superphosphate, as it was called, was the only manufactured fertilizer in use. Of the two nitrogenous fertilizers sulphate of ammonia was only a by-product in the manufacture of gas, and nitrate of soda was simply extracted from the Chilean deposits. There were growing fears, however, that the world's need for fertilizer nitrogen would soon out-strip supplies. Sir William Crookes in his British Association address in 1896 made a very gloomy forecast of the world's future supply of wheat. The wheat lands, he said, would soon all be taken up and the only way of obtaining more food for the growing population would then be to increase yields per acre. Nitrogenous fertilizers would do this, but the visible supplies would soon be inadequate. But he also showed the way out of what he called "this colossal dilemma". Nitric acid could be made from the air by passing it through an electric arc: water power for generating the necessary current was abundantly available and unlimited quantities of nitrate could therefore be made.

This set chemical engineers thinking, and the Scandinavian countries with their vast resources of water power were soon making

synthetic nitrogenous fertilizers. Italy also did the same. This, however, would not have solved the problem completely: production would
have been too localised. Meanwhile the German chemists showed how,
by aid of a catalyst, atmospheric nitrogen and hydrogen prepared
from water could be caused to combine forming ammonia at temperatures far below those of the electric arc, and the ammonia could be
readily oxidised to nitric acid by means of another catalyst. Ammonium
nitrate could thus be prepared in unlimited quantities without the
need for high consumption of electric power: the factories could be set
up anywhere. A high degree of technological efficiency was, however,
required, and for some years Germany alone knew the secrets of the
catalysts; factories were set up there and an unlimited output of ammonium nitrate became possible.

Reference has already been made to the contribution made by the
French military requirements of the 17th and 18th centuries to the
study of nitrification. The First World War had an even greater effect.
Up till 1914 all countries were dependent on Chilean nitrate for the
manufacture of explosives, and in the event of war the British Navy
could cut off supplies. But by 1914 Germany was independent of
Chile and could declare war without fear of this particular trouble.
Great Britain was now in the vulnerable position for we still had to
import nitrate of soda from Chile. After the war British chemists learned
the nature of the catalysts, and Brunner, Mond & Co., later Imperial
Chemical Industries, set up their factory at Billingham; an enormous
output of synthetic sulphate of ammonia and nitrate of ammonia is
now achieved.

The old sulphate of ammonia had been damp, sticky, and bluish
with an unpleasant smell. Usually some sulphuric acid adhered to it
making it unpleasant to handle and liable to rot the bags. It was
somewhat variable in composition and farmers buying large quantities
were advised to have it analysed. The new synthetic material, on the
other hand, is pure, in white crystals, odourless, dry, entirely free from
acid and easy to handle. There has been a similar improvement in the
manufacture of superphosphate: the modern material is incomparably
better than the old. The greatest improvement in modern times,
however, has been in the mixed fertilizers. Before the first war these
were so unsatisfactory that farmers were generally warned against
them. Now, however, they are made of high quality materials, they

are fully up to their guarantee and perfectly mixed; some brands are granulated, each granule containing the complete mixture.

For ordinary garden use a high grade mixed fertilizer is the most effective and convenient, but for farm purposes it may be more economical to suit the fertilizer to the soil or the crop. No analyst, however, can write an exact prescription, and if a mixture is available that approximates sufficiently closely to requirements it could be used. Single fertilizers, are, however, used for special purposes.

Nitrogen is the nutrient most commonly deficient in British soils, and a dressing of a nitrate or an ammonium salt nearly always causes increased growth and a deeper green colour of the foliage. As already stated (p. 69) the ammonium is rapidly changed to nitrate in the soil and in most cases it is probably taken up in this form by the plant. The conversion, however, is not fully quantitative; if the effectiveness of a given quantity of nitrate-nitrogen is put at 100, that of the same quantity of ammonium-nitrogen is about 90: this difference is allowed for in the price. Two other differences are important: the nitrate is easily washed down in the soil, while the ammonium is held by the clay minerals and the humus until it is nitrified. Sulphate of ammonia reacts with calcium carbonate in the soil producing calcium sulphate which, being more soluble than the carbonate, is more easily washed out. If the initial stock of calcium carbonate is low, or if, as on some experimental plots, heavy dressings of sulphate of ammonia are given year after year, the soil may lose all its calcium carbonate and become acid. For many plants this is a disadvantage but not for lawns or golf greens so long as the acidity does not become too pronounced; with this proviso dressings of sulphate of ammonia encourage the fine grasses and discourage clovers. Sulphate of ammonia is well suited for potatoes: they tolerate a certain amount of acidity and it keeps in check the actinomycete that causes scab. About 3 oz. per square yard is a suitable dressing. A mixture of 1 part of sulphate of ammonia with $\frac{1}{3}$ part of ferrous sulphate and 2 parts of dry sand sprinkled on the lawn in dry weather used to be a favourite way of eliminating clovers, daisies and other flat growing weeds; the modern chemical herbicides are more efficient.

For quick action in a backward spring or to aid recovery from the attack of a pest either nitrate of soda or nitrate of ammonia is very effective; the latter is a synthetic product, very concentrated: for convenience of handling it is mixed with fine chalk and sold as nitrochalk. A suitable rate of application is about 2 to 4 cwt. per acre or 1 or 2 oz. per square yard. Some valuable market garden crops such as early cabbage and broccoli in Cornwall may receive much larger dressings.

Nitrate and also ammonium salts because they are so rapidly nitrified must be utilised quickly as they are very liable to be washed out from the soil and hence they are most effectively applied as top dressings when the plant is up. The young plant can absorb a good deal of nitrate and hold it for use later on, but there is a persistent feeling among practical men that a slow but steady supply continuing through all the growing period is better than a quick acting supply given in the early part of the plant's life. This explains much of the popularity of the organic manures described later. A slow acting nitrogenous fertilizer is now manufactured in the United States by condensing urea and formaldehyde; it is known as Uramite, and is much used on golf courses and in gardens but is too expensive for agricultural use.

The most concentrated nitrogenous fertilizer is liquified ammonia which contains no less than 82·4 per cent of nitrogen as compared with 21 per cent in sulphate of ammonia and 16½ per cent in nitrate of soda. It is used to a considerable extent in the United States either dissolved in irrigation water or injected into the soil: it is rapidly absorbed by the clay minerals and the humus. Alternatively it is added to superphosphate which retains it completely. It has not, however, come into use in this country.

Phosphate deficiency is also widespread in British soils especially on the loams and fen soils of the eastern counties, the acid Millstone Grits of the North, and some of the oolite soils of the Midlands and the west. The crops most sensitive to lack of phosphate are swedes, turnips, and potatoes; then come sugar beet, mangold, clover and spring-sown cereals; grass and winter wheat are less sensitive. The effects of phosphatic dressings on farm crops have been studied in much detail: in the seedling stage root growth is hastened, the first leaf is more rapidly passed thus enabling swedes and turnips to escape damage by the flea beetle; later on cereals tiller better, produce more heads and these emerge more rapidly from their ensheathing ears than on plants less well supplied. Phosphates are particularly beneficial to clover on pasture land.

Superphosphate is the most widely used of all phosphatic fertilizers. Its phosphate is soluble in water, it contains also gypsum which is itself a fertilizer supplying sulphur and having beneficial effects on the soil by ensuring a high proportion of calcium among the exchangeable bases (p. 9). It is prepared from calcium phosphate which occurs in immense deposits in North Africa, parts of the United States, and parts of the U.S.S.R.; but in very few other places, and unfortunately the British Commonwealth is particularly badly off. Small local deposits of phosphate rock are not uncommon, but they often contain iron and aluminium compounds which make them unsuitable for the manufacture of superphosphate. This very restricted distribution of the raw material for making phosphatic fertilizer may lead to difficulties when the undeveloped countries begin to intensify their food production in order to raise the level of nutrition of their undernourished people. Immense supplies of phosphatic fertilizer will be needed: nothing else will take its place. Superphosphate should be applied at the time of sowing or planting, and placed near to the seed so that its full effect on root development can be exercised; 2 to 4 oz. per square yard is ample for gardens. It used to be said that superphosphate made the soil acid, but this is incorrect.

In view of the fact that the sulphur required for making superphosphate by the usual process has to be imported at considerable cost various other methods of making the rock phosphate available have been tried. None gives a soluble phosphate; some of the products are useful on acid soils but none is as effective as superphosphate on neutral soils.

Before the time of superphosphate ground bones were the only phosphatic manure available though it was not known that their value was due to the phosphorus. Bone phosphate is not soluble in pure water, but it dissolves sufficiently in the water of acid soils to be taken up readily by plants. Ground bones and bone meal usually contain between $3\frac{1}{2}$ and 4 per cent of nitrogen which adds to their value: the phosphate (20 to 25 per cent of P_2O_5) has about the same fertilizer effect as that in superphosphate and many gardeners prefer it on the grounds that it cannot easily be misused, which is true, and that its effects are more lasting, which is not true. 2 to 4 ozs. per square yard are sufficient but an excess does no harm—nor any good. In planting rose bushes a favourite device is to put some peat moss litter and bone meal round the roots; both encourage development of a vigorous system of rootlets.

Another bone manure, steamed bone flour, is richer in phosphate but poorer in nitrogen than bone meal: it is also more finely ground. These differences arise from the fact that in making bone meal only the fat is extracted from the bones, while steamed bone flour is a by-product of the glue industry for which the protein of the bone is extracted. Steamed bone flour is more economical if the phosphate effect only is required and the area to be fertilized is large, but for ordinary garden use bone meal has the advantage of supplying also some nitrogen in a very available form.

For farm use on grazing land high grade basic slag has proved a valuable fertilizer, encouraging the growth of wild white clover which greatly improves the herbage. Modern slags contain a proportion of their phosphorus in the unavailable form of a fluor-apatite; they should be purchased only on their content of phosphate soluble by the standard citric acid test.

Whichever of these fertilizers is used the phosphate becomes insoluble in water in the soil and plants take up only about 25 to 30 per cent of the quantity added even over a run of years. The rest stays in the surface soil apparently for ever: almost inert, but without losing its solubility in dilute acids which distinguishes it from the original phosphate of the soil. Sites once inhabited and now long since abandoned can still be distinguished by their higher content of this soluble phosphate compared with that of surrounding land—the residue of household waste, ordure, bones cast out in the days before public health services existed. F. Hughes observed this in 1911 in old centres of habitation in Egypt, and ten years later O. Arrhenius recognised in this way Mesolithic sites thousands of years old in Sweden; more recently the Archaeology Department of the Birmingham University traced the boundaries of a Roman Station at Brough, the Roman posting station Crococalana lying across the Fosse Way in Nottinghamshire. Similarly the sites of abandoned medieval villages were recognised.[1]

This fixation of phosphate is most marked in very acid or alkaline soils; addition of lime therefore increases the availability of the soil phosphate if the pH value is below about 6.5 but decreases it if the pH is 7 or more.

When superphosphate made from radioactive phosphorus (P^{32}) is added to the soil some of its phosphate ions exchange with those of the soil with the result that the additional quantity taken up by the plant

[1] Described by K. D. M. Daunsey, *Adv. Sci.* 1952, 9. 33–36.

has come partly from the soil and not wholly from the superphosphate as had previously been supposed.

Potassium is much more widely distributed in soils than phosphorus. The various forms in which it occurs are described on pages 11 and 15: clay soils usually are best provided and sandy soils least. The soil potassium is most easily available on acid soils: indeed in very acid conditions (pH 4·5 or less) it is liable to be washed out; it is least available at pH values above 7·2, e.g. on chalk soils.

Potassic fertilizers came much later into use than the phosphatic and nitrogenous; nineteenth century farmers hardly knew them. The old system of agriculture and the Rules of Good Husbandry that dominated the leases conserved the supplies of potassium in the soil, and the crops that most need it: mangolds, sugar beet and potatoes, were not as yet much grown. Hay and straw both contain considerable quantities of potassium: their sale off the farm was generally forbidden to tenant farmers; they were consumed by the farm animals and most of the potassium got into the excreta (chiefly the urine) and was returned to the land in the farmyard manure which was then usually richer than it is now. Changes in agricultural systems and complete removal of all restrictions on sales of produce have drastically altered the position. Crops needing potassium are much more widely grown than before, yields are considerably higher, and owing to the high cost of labour conservation is no longer as effective as in the old days. The result is a great increase in the consumption of potassic fertilizers. Fortunately the raw materials are available in immense quantities, and vast deposits are more widely distributed over the world than the rock phosphates.

Potassium increases the efficiency of the leaf in making sugar; it is thus an effective companion to nitrogen which increases the area of the leaf. On the Rothamsted mangold field one ton of leaf produced 4·3 tons of roots when potassic fertilizer was supplied, but only 2·9 tons when none was given. Potassium also increases the health and vigour of the plant enabling it better to resist certain fungus diseases and to stand up to drought and other adverse conditions: at Rothamsted the foliage of potatoes supplied with potassic fertilizer continued growth in a dry season long after that of plants without it was already dead. Potassium strengthens the straw of cereals and of grasses and increases the proportion of leguminous plants in hay. It also improves the quality of fruit and the cooking quality of potatoes, and is thus of

a Magnesium deficiency: apple leaves.

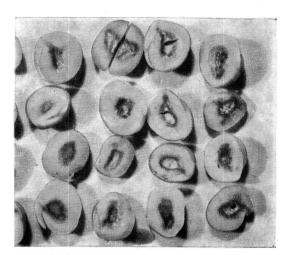

b. Manganese deficiency: pea seeds.

Plate 2. Symptoms produced by deficiency of
a. magnesium, *b*. manganese (*by permission of the
Controller of H.M. Stationary Office*)

special importance to market gardeners: much is used in the Vale of Evesham and other fruit and vegetable growing regions.

Light chalk soils, sandy and moorland soils respond particularly to potassic fertilizers.

The two most widely used are the sulphate and the chloride, commonly called the muriate; the sulphate is more usual in gardens and the muriate, being cheaper, on farms. Both are equally good so far as the potassium is concerned, but the chlorine in the muriate is liable to harm good quality potatoes and possibly other market garden crops also; the sulphate ion on the other hand has no injurious effect and indeed the sulphur may often be beneficial. Sulphate of potash is much used for high quality produce. Either fertilizer can be applied at the rate of 2 cwt. or more per acre or 1 oz. per square yard. One ton of farmyard manure supplies potassium equivalent to about 22 lb. of muriate or sulphate of potash.

Marked deficiency of potassium is shown by dying back of the tips of shoots and leaves and by chlorosis of the leaves: they also die at the edges: Pershore plums show this markedly. Apples and bush fruit suffer from leaf scorch. Plants differ in their susceptibility: instances have occurred in Ireland of potatoes showing serious symptoms of potassium deficiency, yet the succeeding oat crop did not, nor even did it respond to potassic fertilizers. Some of the trouble may be due to a high level of other elements, notably nitrogen and phosphorus in the cell, consequent on the low level of potassium. The ultimate cause, however, at least in the case of barley, seems to be an accumulation of putrescine, a product of the disturbed metabolism when supplies of potassium are low.[1]

Although sodium is not essential for plant growth it acts beneficially in certain cases. To some extent it can replace potassium, but some crops respond to dressings of agricultural salt even when ample supplies of potassium are present: chief among these are celery, turnips, table beet, and asparagus; and among farm crops, sugar beet and mangolds.

Changes in the character of modern farming, and the need for intensifying output, have combined to increase greatly the draft on the stocks of plant food in the soil: G. W. Cooke's estimate of the losses of plant nutrients per acre on the old conservative farming systems compared with present day cash cropping are given in Table 13.

[1] F. J. Richards & R. G. Coleman, *Nature*, (1952) 170. 460.
WOS—L

TABLE 13
Nutrients removed in crops in the four year period, lb. per acre:

	N.	P₂O₅	K₂O
Old four course rotation	57	18	7
Modern systems ..	390	128	435

From *Agricultural Progress* (1954) 29, 110–120; a short paper with much valuable information on fertilizer practice.

The need to replenish and even to increase these stocks has led to an enormous increase in the consumption of fertilizers in advanced countries as shown in Table 14.

TABLE 14
Quantities of fertilizers used by farmers in the United Kingdom, expressed in thousands of tons of plant food, in various years:

	NITROGEN (N)	PHOSPHORIC ACID (P₂O₅)	POTASH (K₂O)
1900	16	110	7
1913	29	180	23
1929	48	198	52
1939	60	170	75
1946	165	358	120
1956	291	386	305

Large as the modern quantities are, the Ministry of Agriculture does not consider the limit is anything like reached: in 1952 it was estimated that consumption of nitrogen could advantageously be doubled, while that of phosphoric acid could be raised by one third and of potash by one half.

Besides these three classical nutrients there are about ten others equally necessary to the plant: usually the soil contains enough but not always, and deficiencies may occur which are often the cause of considerable trouble becoming more pronounced as the intensity of farming increases; they have been much investigated by T. Wallace and his colleagues at Long Ashton. The plant responds to a deficiency by developing characteristic symptoms[1] by which it may be recognised, but chemical analysis of the leaves may be needed for complete diagnosis. The effects are not always due to an actual lack of the element

[1] Described and illustrated in his book *The diagnosis of mineral deficiencies in plants* (1951), 2nd Edn. London, H.M.S.O.

in the soil: they may result from its unavailability or from a lack of balance with other elements in the plant.

Magnesium is an essential constituent of chlorophyll, the green colouring matter of leaves and stems, and it is also concerned in the phosphate nutrition of plants. Magnesium deficiency shows itself in the leaves: parts of them may lose their colour and become chlorotic, or they may turn various shades of red or purple; the result may be blotchy but is often a regular and attractive pattern varying somewhat with the plant. The effects are especially marked in later stages of growth. Magnesium deficiency is fairly common in horticultural plants both in gardens and orchards, and is at present perhaps the most serious deficiency from which apple trees suffer. It is especially liable to occur in intensively managed glass houses: some of the most beautiful examples I ever saw were on tomatoes in the glass houses of Westland in the Netherlands, one of the most highly cultivated regions in the world. In outdoor crops there it is more common in wet springs than in dry springs. (Plate 2, p. 146).

Magnesium deficiency is often induced by a high potassium and low nitrogen content of the plant; it is likely to increase as the consumption of potassic fertilizers increases—which it must do in order to provide the increasing amounts of food required by an increasing world population. The obvious remedy of adding magnesium salts to the soil may be ineffective if much potassium is present. Fortunately, however, mineral nutrients can enter the plant by the leaf as well as by the root, and a solution of 2 per cent magnesium sulphate sprayed on to the young foliage supplies the plant with magnesium ions without the complicating presence of potassium ions. Deficiency of magnesium in the blood of cows and sheep causes the trouble known as hypomagnesaemic tetany first studied in Holland in the late 1920's and a few years later in Britain. It seems to be induced by heavy dressings of fertilizers, especially sulphate of ammonia, to the pastures but is not directly correlated with the magnesium content of the herbage. Reports of cases are increasing especially among calves and sheep: it is not known, however, to what extent this is due to better recognition of the symptoms.[1]

Calcium deficiency is generally associated only with very acid soils and there the effects are complicated by other actions of the acidity. The typical symptoms are dying back at growing points,

[1] Ruth Allcroft, (1954) *Veterinary Record*, 66, 517.

wilting of the stems and flower stalks, defective root systems, and failure to produce tubers.

Deficiency of sulphur has not been observed in Britain, partly because sulphur is contained in most of the fertilizers used, and partly because considerable quantities of sulphuric acid are poured into the atmosphere from household fires and industrial operations and are brought down in the rain. In Continental regions where fertilizers are not used and the atmosphere is purer sulphur deficiency is not uncommon: cotton and lucerne frequently suffer in the United States, and some interesting cases are reported from Africa where not only tea and other plants are affected but even fishes in the lakes.

The remaining six elements are needed in minute quantities only; they are called "trace elements". Two have been recognised for a long time: iron and manganese. Iron is essential for the proper functioning of the chlorophyll; deficiency shows itself by the leaves turning light green to yellow between the veins which, however, remain dark green; in severe cases the whole leaf turns yellow and dies; shoots and branches also die. Some plants such as gardenias, azaleas and rhododendrons are so sensitive to iron deficiency that they cannot easily be grown on any but acid soils, for it is only in these that there is usually sufficient soluble iron to serve their needs. On calcareous soils, on the other hand, the calcium carbonate not only throws the iron out of solution but actually prevents the plant utilising such iron as it is able to pick up. Lime-induced chlorosis is for this reason very difficult to rectify. There is no lack of iron in the soil: indeed there is often as much as in non-calcareous loams and more than in some of the sands.

A number of causes may prevent the plant from obtaining iron from the soil even when it is present in quantity: among them are toxic amounts of copper, zinc or other heavy metals, or other adverse conditions in the soil, nutritional disturbances in the plant, and lack of potassium. Iron deficiency cannot be overcome by adding more iron salts to the soil, nor as a rule by spraying the plant with ferrous sulphate, though fruit trees may benefit from crystals placed in a hole bored in the trunk. Some varieties of fruit trees are less susceptible to iron deficiency than others and the trouble may be evaded by growing these. Fruit trees on the calcareous soils of eastern England are particularly liable to be affected, though the trouble can be mitigated or even overcome by growing grass round them. Ordinary farm crops are less liable to suffer. (Plate 1, p. 18).

A promising method of dealing with iron deficiency has been developed in Florida. The citrus fruit crop is of great importance there, but more than half of the trees were reported to be suffering from iron deficiency. Addition of ferrous sulphate to the soil was impracticable: up to 200 lb. per tree were needed to effect a cure. Advantage was taken of the property possessed by some organic compounds—to which reference has already been made (p. 16)—of being able to dissolve iron oxide from the soil and convert it into a form in which it can pass through the soil without being absorbed by the colloids. Ordinary iron salts cannot do this, and the relationship between the organic compound and the iron is complex and different from that of an acid and a base forming a salt, the iron being in a non-ionised condition. Some of these solutions can pass from the soil into the plant, getting the iron safely in and leaving it to do its work there. The organic compound is given the name chelate,[1] from the Greek word χηλή, a claw; because of the calliper-like action of two of the groups in th molecule which fasten on to the metallic atom and bring it into a heterocyclic ring, the idea being that the chelate can get hold of the iron atoms putting them into a combination from which they cannot dissociate: in consequence they behave quite differently from those in ordinary iron salts.

A number of compounds are known that can act as chelates, but in order to be of practical value they must resist fixation or decomposition in the soils long enough to enable plants to absorb them. This requirement is not easily satisfied because, as we have seen, the soil microorganisms can break down a surprising number of organic compounds. Fortifying with chlorine atoms is not permissible because that might make the substance toxic to the plant. Success was achieved by using an ethylene derivative, ethylene diamine tetracetic acid, called EDTA for short. This substance is stable in the soils of Florida—which are mostly non-calcareous—and remains therein undecomposed for months, at the same time it is harmless to plants in the quantities used. Its combination with iron enters the plant root, the plant chelates take the iron and distribute it to the places where it is needed. The process is so efficient that only small amounts of iron are required. In place of the 200 lb. per tree of ferrous sulphate needed when it was added to the soil, a quantity equivalent to $\frac{2}{3}$ oz. was adequate; in some cases even

---

[1] The name was given by their discoverer, G. T. Morgan (*J. Chem. Soc.* (1920), 117, 1456 and (1925) 2030).

less. The response was immediate. Citrus trees so chlorotic that they produced no fruit at all were stated to have yielded 600 lb. in the same year in which they had been treated; roses improved enormously and other garden plants showed marked benefit. Azaleas, gardenias and rhododendrons that usually will not thrive on calcareous soils did so after treatment with the chelate. It has proved so successful in Florida that it has come into commercial use; the sodium salt is mixed with the fertilizer for ease of application.

Unfortunately it is not very effective on calcareous soils, and it is just on these that the trouble is worst in England. The chelate is decomposed in the soil. Experiments at Long Ashton, however, show that it gives good results as a foliage spray. The present chelates represent only the beginning: now that a pattern has been found there are many possible variations, and already one of them in which cyclohexane replaces ethylene as the scaffold is an improvement. As with other agents, the action of chelates is not entirely simple. They act also as growth promoting substances (not however, very potent) and when used as sprays soften the water by making the calcium non-basic.

Manganese in small quantities is necessary for respiration and for protein synthesis, being a constituent of the enzymes concerned in these processes. As in the case of iron, deficiency is common on calcareous soils and can occur even where abundant supplies are present, the reason being that only divalent manganese appears to be effective, while the quadrivalent form to which it is readily transformed by oxidation is not. The divalent form is more stable in acid than in neutral soils, and the cultivator is often placed in the quandary that unless he limes his soil the plants will suffer from acidity, but if he limes it they suffer from manganese deficiency. The trouble is particularly liable to occur in conditions of impeded drainage, presence of calcium carbonate, and of blackish organic matter.

The symptoms vary in different plants; the leaves usually become chlorotic, i.e. lose their green colour, but the effect is generally patchy or spotty. In oats the trouble is called Grey Speck disease although actually the chlorotic areas are often in the form of narrow stripes along the leaf as well as specks; in peas it is called Marsh Spot, and in sugar beet, Speckled Yellows. Leaves of fruit trees become yellowish at first near the margins and later between the veins towards the midrib, but the veins and the adjoining tissue retain their green colour; the symptoms are not unlike those of iron deficiency. Plants growing on calcareous soils or on newly limed peat soils are particularly liable to suffer. (Plate 2, p. 146).

Of all this group of elements manganese appears to be most frequently lacking from the soil; practically all farm and garden crops are affected in consequence. Addition of manganese salts to the soil is not usually effective; a better remedy is to spray the plants with a solution of manganese sulphate at the rate of about 3 lb. of the salt in 100 gallons of water per acre for fruit trees and garden plants: higher rates are sometimes used on farm crops.

While a certain degree of soil acidity increases the availability of the manganese a high degree may bring so much into solution as to be toxic to plants. Waterlogged conditions may do the same as also may sterilising the soil by heat.

The other four trace elements are of special interest because although absolutely essential the quantities needed are so minute that only in recent years have improved analytical methods shown their significance. They further differ from the classical nutritive elements in that even small quantities in excess of plant requirements may cause serious damage to the plant: the line between benefit and injury is easily overstepped and these elements should be applied only under competent advice.

Boron was one of the first of this group to be studied, and as has often happened the discovery was accidental. The entomologists at Rothamsted were trying to protect broad beans against black aphis and among the methods tested was the addition of various salts to the soil which if absorbed by the beans might make the leaves distasteful to the aphis. Not knowing exactly what the tastes of the aphis might be a large number of substances had to be tried. The experiment failed in its primary purpose, none of the substances protected the beans. But one of them, borax, increased the growth of the plant. The information was passed on to the Botanical Department and experiments showed that the elements previously regarded as sufficient for the growth of the bean were in reality not so; a small quantity of boron was necessary: it had been overlooked because the nutrient salts ordinarily used in botanical laboratories contained a sufficient amount as undetected impurity. Detailed studies showed that boron is essential to all plants; without a sufficiency calcium uptake and utilisation are impeded, and the meristematic tissues fail to differentiate, resulting in deformed growths. Root crops and brassicas are especially liable to be affected. Beet and turnips suffer from "Heart Rot" and "Canker", celery from "Cracked Stem", apples from "Corky Core", or "Corky Pit". A special effect on beans is described on page 65. (Plate IXa).

Boron deficiency is common on light sands and gravels under high rainfall, also on soils derived from crystalline rocks and fresh water sediments: it is most marked in dry summers on sandy soils containing much calcium carbonate but little clay or organic matter. 1955 was a notoriously bad season. It can be dealt with by incorporating a small amount of borax—about 20 lb. per acre—in the fertilizer and by using as nitrogenous fertilizer Chilean nitrate of soda, which contains a sufficient, but not a harmful amount.

The importance of copper was first demonstrated on some of the newly reclaimed peat and heather moor soils of West Germany and the Netherlands. Cereals would not grow well, although on older cultivated peats they flourished. The name "Reclamation disease" (*Urbarmachungkrankheit*) was given but no remedy was known. On one occasion Professor Hudig of the Groningen Experiment Station in the Netherlands was visiting the Moor Culture Station at Bremen and observed that the cereals growing on the high moor peat soi's suffered in the same wayas those in his own country except on one plot where they were healthy. The German investigators attributed this to a higher resistance to night frost, then regarded as the cause of the trouble. Professor Hudig, however, was not satisfied: his further enquiries revealed that a previous crop on the healthy plot had been treated with a copper spray and on returning home he started experiments to find out whether copper sulphate would cure reclamation disease. It did, and copper sulphate proved to be a suitable remedy elsewhere; reclamation disease is now completely under control. The trouble was not at an end, however: oats sown after the potatoes sometimes suffered from Grey Speck disease, and this could not be cured as easily as usual by application of manganese salts; the copper had somehow counteracted the effect of the manganese, presumably by stimulating the reaction which renders manganese unavailable to the plants.

Copper is essential for plant growth being a constituent of at least one of the enzymes concerned with oxidation. (Plate X, p. 163). Deficiency causes dying back of the growing points, deformed growths, gumming in some fruit trees, the leaves may be chlorotic or bluish green. The quantity of copper required is extremely small—some 3 to

5 parts per million of weight of dry matter of the leaf of the apple tree suffices—and only in the last twenty-five years has its necessity been recognised.[1] Copper deficiency is now known to be widespread in Wales (where most of the soils are formed from very ancient rocks), it affects apples and pears on sandy soils in southern England, and cereals on the reclaimed fen and sandy heath soils in the Eastern Counties, particularly the peat fens. The old process of marling, i.e. digging down to the under-lying bed of clay and bringing some of it up for spreading on the surface, is a satisfactory remedy where it is still practicable; alternatively a dressing of about 20 lb. per acre of copper sulphate can be given or the crop can be sprayed with a 0·1 per cent solution. Less than 1 lb. per acre of copper sulphate so applied has raised the yield of a cereal crop from 2 or 3 cwt. per acre up to 30 cwt. per acre. Much higher dressings—45 or even 90 lb. per acre—are required on newly reclaimed heath and peat soils in the Netherlands. Crops on reclaimed bog land in Eire also respond remarkably to dressings of copper sulphate. Copper oxychloride is said to be equally effective and safer in use.

Sheep feeding on pastures deficient in copper are liable to produce abnormal wool; a deterioration of the "crimp" is particularly noticeable in Australia among Merinos, and the wool is described as "steely".[2] In Great Britain the lambs born of ewes on such pasture suffer from "sway back".

The necessity for zinc was discovered independently by two groups of United States investigators: in both cases through shrewd following up of an accidental observation. Apple trees in California suffered from "little leaf", mottling, meagre growth, shortened internodes producing rosette appearance, and dying back of the growing tips. Citrus trees also suffered similarly. The first experiments suggested that spraying with a solution of ferrous sulphate was a satisfactory remedy. When tried on the commercial scale it appeared to succeed in small orchards where everything had to be done cheaply but not in the large ones where better materials could be used. Investigation showed that only the cheaper impure grades of ferrous sulphate were effective, not the more expensive purer salt. Among the impurities was zinc, and this

[1] It was first demonstrated in water cultures in 1931 (*Plant Physiol.* 6, 339 (A. L. Sommer, & 593, C. B. Lipman & G. Mackinney).

[2] Copper is a constituent of one of the oxidases needed for the folding of the keratin molecule to produce the normal structure of wool. Metabolic disturbances are likely to occur in its absence. It is also needed by slugs and snails (p. 131).

proved to be the active agent. The other group of investigators obtained good results with ferrous sulphate in their small scale experiments, but growers on the large scale did not. Examination showed that the investigators had used galvanised iron vessels for holding the solution while the growers used wooden barrels; sufficient zinc had been dissolved from the galvanised iron to cure the disease.

The amounts required by the plant are larger than for copper, though still very small: 10 to 40 parts per million of the dry matter of the plant. Deficiencies have only recently been recognised in Britain. A particularly interesting example occurs at Wisley in the gardens of the Royal Horticultural Society where apple trees are affected.[1] W. A. Roach of East Malling has reported cases on a Kentish marsh soil. The availability of soil supplies of zinc and of copper, like that of manganese, is decreased by calcium carbonate and in some cases by a high content of organic matter.

Molybdenum is remarkable as being the only element of high atomic weight essential for the growth of plants. The amount needed is extremely small,—1 part per million of dry matter is said to suffice for clovers—and the discovery was possible only because of the increasing refinement of chemical technique. It was in Australia that the discovery was first made: on some of the sandy soils subterranean clover, a highly important crop in the southern regions, would not grow without a dressing of $\frac{1}{2}$ to 2 lb. per acre of sodium or ammonium molybdate. Similar results have been obtained in parts of New Zealand. (Plate IX*b*, p. 162). Investigations elsewhere showed that tomatoes, lettuce, cauliflower and broccoli, rye grass and other crops needed it and there seems little doubt that this is true of all plants. It is also necessary for the nitrogen fixing organisms. Deficiency of molybdenum is shown by chlorosis of the leaf tissues and inability of the laminae to develop, resulting in such peculiar growths as "whip tail" of cauliflowers. T. Wallace states that as little as 2 to 4 oz. per acre of molybdate will supply the needs of the cauliflower crop. Troubles due to lack of molybdenum are most commonly observed on acid sands newly brought into cultivation: they are intensified by soil acidity and excess of manganese. They can often be rectified by dressings of lime, but on some soils addition of sodium molybdate may be necessary. It has been claimed that the effect of lime in some cases is mainly to bring unavailable molybdenum into action and that a small

[1] J. M. S. Potter (1953) *J. Royal Hort. Soc.* 28, 260-7.

molybdenum dressing can take the place of a large dressing of lime.

Some of the Lower Lias clay soils of Somerset contain so much molybdenum—more than 300 parts per million compared with the usual 0·2 to 5 parts—that the grass is injurious to calves and young cattle, causing them to scour badly: there are called "teart" pastures: fortunately they are only of local occurrence.[1]

Diagnosis becomes more difficult when two or more of the trace elements are deficient simultaneously, so confusing or even concealing the usual symptoms: deficiency of manganese may be accompanied by deficiency of copper or zinc, and excess may cause a deficiency of iron on strongly acid soils. Roach found potatoes suffering from manganese deficiency on a Kentish marsh soil; spraying with manganese sulphate removed the symptoms but did not improve the yield. Zinc deficiency was suspected, and spraying with a mixture of zinc and manganese sulphates put up the yield considerably.

Deficiencies of these trace elements are widespread in Australia and not uncommon in New Zealand; some notable reclamations have been effected in both countries by first discovering which are lacking and then supplying them. One of the most remarkable of these enterprises is the conversion of a vast area of former waste land once known as the Ninety Mile Desert between Melbourne and Sydney into a productive dairy farming region. Important and interesting investigations on the subject have been made at Adelaide both by Dr. Marsden and his colleagues, and at the Waite Institute: some of C. S. Piper's results are shown in Plate X (p. 163).

Cobalt is needed in minute quantities in the nodules of leguminous plants, for without it they will not fix nitrogen. It is also needed in grass, for animals need it as it enters into the molecule of Vitamin B_{12}.[2] Iodine is apparently not needed by plants although it also is essential for animals.

From the foregoing paragraphs it appears that plants require thirteen elements from the soil, six in relatively large amounts and the rest in much smaller quantities: they are:—

[1] The affected herbage may contain 20 to 100 parts of molybdenum per million of dry matter while normal herbage contains only about 0.5 to 1.5 parts. Yorkshire Fog takes up most, clover less, other grasses still less. The harmful effect on the cattle increases as the proportion of clover in the pasture rises. The effect can, however, be counteracted by dosing the animal, with 1 to 2 grams of copper sulphate daily (W. S. Ferguson, A. H. Lewis and S. J. Watson, *J. Agric. Sci.* (1943), 33, 44.)

[2] See D. P. Cuthbertson and Ruth Allcroft for an account of trace elements in animal nutrition (Advancement of Science (1956), 12, 485–497).

Large amounts: nitrogen, phosphorus, potassium, calcium, magnesium, sulphur;

Small amounts: iron, manganese, boron, chlorine, copper, zinc,
molybdenum. Silicon appears to be essential for rice and perhaps
for some other cereals.

Not essential but nevertheless beneficial at least to some crops:
sodium, especially for the beet family.

The suitability of the nutrient supply depends not only on the
absolute quantity of the individual ions in the plant but also on their
relative proportions. An excess of calcium may induce deficiency effects
of potassium, iron, magnesium, manganese and boron, even though
these substances may be present in amounts that would otherwise be
adequate. A proper balance between the various elements is essential,
particularly in conditions of intensive cultivation.

There may be some interaction between roots and solid particles
in the soil, but most of the substances entering the plant come from the
soil water. This, however, does not enter the plant *en masse* in the
mechanical way in which it would enter a drain pipe. The plant has
some power of selection. Some ions are taken in relatively large
quantities, others in much smaller amounts; the proportions of the
different ions finally present in the plant differ considerably from those
in the soil solution. The plant has no power of complete rejection of an
ion; it may take up a poisonous substance and be killed thereby. On
acid soils brassicas may suffer from taking up too much manganese,
and beetroot and celery from too much aluminium; harmful excess of
chloride may be absorbed by red currant bushes, potatoes and other
plants. On the other hand some plants have difficulty in getting
sufficient of certain essential ions: turnips and potatoes often cannot
pick up enough phosphate even though grasses, cereals and other
plants growing alongside of them acquire all they need. But plants can
accumulate ions. Leguminous plants accumulate calcium; cereals and
grasses accumulate silica; *Atriplex hortensis*, various sea-side and other
plants including chickweed accumulate sodium; more than 2,000
plants are known that accumulate aluminium,[1] while some plants,
e.g. *Carex humilis* and *Arabis stricta*, apparently accumulate strontium[2].
Nemec states that in Bohemia *Equisetum palustre* accumulates gold. It
is not clear why certain ions should be piled up in this way: cereals

[1] *Rothamsted Annual Reports*, 1948, p. 35 and 1949, p. 40.
[2] H. J. M. Bowen and J. A. Dymond (1955) *Proc. Roy. Soc.* 144B, 355-368.

and grasses can get on perfectly well without silica when they are grown in artificial media free from it, nor have aluminium or gold ever been shown to be necessary for plant growth.

The process of entry of ions into the plant is still the subject of discussion. According to the older view it occurred in two stages: adsorption of the ion on the root surface, and its transfer into the plant system. Some ions interfere with the adsorption of others: it was supposed that a particular group of ions can be adsorbed on particular sites only and not on those appropriate to other groups; that within the groups there is antagonism between the different members; if one is adsorbed another cannot be: potassium for example depresses the entry of magnesium into the plant. Some of the adsorbed ions are then carried in to the plant system by certain molecules in something like a chelating action. Later workers (whose views are by no means universally accepted) have given up the idea of a selective membrane at the external surface of the root cells, and suppose that ions get freely in and out of the root by simple diffusion. Some of the ions are bound in the cytoplasm, but they can exchange with other ions in the incoming solution. This region of the root, roughly the cell walls and the cytoplasm, is called the free space: it is immediately accessible to ions from outside. On the inner surface of the cytoplasm bounding the vacuole there is a selective membrane and it is here that the process of selection goes on.

However it is effected the final accumulation of ions within the plant is not a simple mechanical in-flowing. It rather resembles pumping in that it necessitates the expenditure of energy and this comes from the oxidation of sugar in the root by oxygen from the soil air at the surface of the root. Any slowing down of this oxidation process, e.g. by waterlogging, interferes with the absorption of the ions and therefore with the nutrition of the plant. The concentration of bacterial population round the root might be expected to reduce the amount of oxygen and increase that of carbon dioxide both of which effects would be harmful: about this, however, nothing can be done.

An adequate supply of soil moisture is equally necessary to ensure mobility of the ions.

While it is essential for detailed investigation of plant nutrition to isolate the problems and study each individually it is equally essential for a full understanding of the process to realise its intimate association with the air and water supply of the soil and therefore with the soil structure, which is closely related to the amount of organic matter in the soil.

CONTROL OF THE SOIL ORGANIC MATTER

Composts and Farmyard Manure

Long before the dawn of history it was known that human and animal excrements greatly increased the growth of plants and they were regularly applied to the land for this purpose. The making of compost was well developed by the eastern peoples, and in his *Book of Agriculture* written in the 12th Century Ibn-al-Awam, a Moor living in Seville, gives details of how they should be made. A pit should be dug, excrements of human beings, doves (from dove cotes) asses and cows should be put in, but human excrement should predominate; an equal quantity of powdered earth from under manure heaps should be added, then blood: human blood is best, he states, but camel and sheep's blood is also good; then water should be added and the whole well mixed. If rain comes it activates the decomposition. When the mass has turned black it should be dried, mixed with more powdered earth and applied to the small garden plants: mint, endive, basil and estragon (an Artemisia: both were herbs used for flavouring), beet, cress and winter cress, purslane, parsley, are all specifically mentioned. Apart from drying which would cause loss of ammonia, minimised however by the considerable volume of soil, there is little one could add to this recipe and undoubtedly the product would be effective.

Most of these materials are no longer available in any quantity and the modern compost heap is perforce mainly vegetable containing a rather high proportion of carbon in relation to its nitrogen. Partial decomposition is essential before it can generally be used as manure. As has already been pointed out the micro-organisms bringing about this decomposition are greatly limited by the shortage of nitrogen and phosphate, and the secret of making good compost is to supply these and other requirements of the micro-organisms.

The nitrogen requirement was determined by H. B. Hutchinson and E. H. Richards at Rothamsted: they found that 0·75 parts of nitrogen should be added per 100 of straw, roughly 6lb. of sulphate of ammonia per cwt. of straw. Garden refuse, especially young plant material, is, however, usually richer in nitrogen than straw, and may require only 1 to 2 lb. of the sulphate per cwt. Some phosphate is usually necessary also. Suitable "starters" are on the market with instructions for use. The heap must be kept sufficiently moist and aerated and is improved by accessions of nitrogenous material; bed-room water, household wastes and the like. Some garden refuse, such as early lawn mowings, contain more than enough nitrogen and can be put straight on to rose beds or around fruit bushes as a mulch and afterwards worked in to the soil; the excess of nitrogen over and above what the organisms require is converted first to ammonia and then to nitrate which the plants can take up. But if older material poor in nitrogen is worked direct into the soil the micro-organisms take up some of the soil ammonia and nitrate, thus depriving the plant of them. Professor R. H. Stoughton of Reading, in his studies of the possibility of composting sewage sludge and household waste, put the critical ratio of available carbon to available nitrogen at about 20: at lower ratios nitrogen is liberated and at higher ratios it is immobilised. Badly made compost, straw and other materials poor in nitrogen applied in spring can do much harm, not only by robbing the plant of available nitrogen but also by opening up the soil so that its moisture can evaporate: the harmful effect is intensified by the slowness with which the material decomposes. In autumn the case is different: undecomposed compost may do no harm, and may even be beneficial by facilitating drainage on a heavy soil and by fixing the nitrates.

Organic matter plays such an important part in the soil that none should be wasted if it can be converted into humus in the soil. But some breaks down too slowly: wood, sawdust, paper are useless except for burning; wood leaves an ash containing about 10 per cent of potash (K_2O)—about one-fifth of the amount in the sulphate or muriate. All diseased plant material should in any case be burnt.

Composts suffer from the weakness that they contain only the mineral ingredients taken from the soil; any deficiency in the soil is reflected in the compost heap. If the deficiency is recognised it can of course be rectified by adding the appropriate fertilizer, but if it is not, it is likely to be perpetuated where compost is the only soil amendment.

Farmyard manure is less liable to this particular trouble. It is virtually a compost made up of two components, animal excretions and straw, but the composition of the excretions is related to that of the food, and an important fraction of this has been brought in from several other districts, some of them overseas. The animal adds no new material and effects no improvement; the changes effected within its body do no more than micro-organisms could effect if the food and straw were made into a compost heap and kept moist. Indeed there is a good deal of loss, for some of the plant nutrients are retained by the animal or pass into the milk, and much of the urine, which contains much of the nitrogen and the potash, fails to be collected under modern methods of management: farmyard manure as made on a good farm eighty or more years ago was undoubtedly better than today's product. But it still remains one of the most useful manures available to farmer or gardener. One ton contains as a rule some 12 to 16 lb. of nitrogen (N), 5 to 9 lb. of phosphoric acid (P_2O_5) and 12 lb. of potash (K_2O); 10 tons of farmyard manure per acre (a common dressing) supply about the same quantity of nutrients as a dressing of 5 to 6 cwt. sulphate of ammonia, 4 to 6 cwt. of superphosphate and $2\frac{1}{2}$ cwt. sulphate of potash. This mixture if supplied as fertilizers would be regarded as very ill-balanced, the nitrogen being excessive in relation to the phosphate and the potash. Actually however, the nutrients in farmyard manure have little more than about half the value of those in the fertilizers, even allowing for the residual effects. Field experiments show that in good conditions crops take up about 50 to 65 per cent of the nitrogen supplied in sulphate of ammonia but only 25 to 40 per cent of that in farmyard manure; the uptake of phosphoric acid from fertilizers is about 25 to 30 per cent but from farmyard manure is less; there seems to be little difference, however, in the uptake of potassium. The advantage of farmyard manure, like that of other composts, is that it supplies organic matter to the soil.

It is not unusual to give responsive crops like sugar beet or potatoes a dressing of farmyard manure in the autumn and of a mixed fertilizer in spring. A curious result is then obtained. The farmyard manure depresses the action of the potassic fertilizer, slightly depresses that of phosphate, but has no action on the nitrogenous fertilizer so that each unit of nitrogenous nutrient exerted its full effect.

Dressings of farmyard manure increase the amount of organic matter and of nitrogen in the soil but with a steadily increasing loss as the dressings continue. This is clearly shown on the Broadbalk wheatfield at Rothamsted where one plot has received 14 tons per acre of farmyard manure annually since 1843 and its neighbour receives no manure of any kind. Over the first 22 years the net increase of organic matter in the soil was 22 per cent of the total quantity

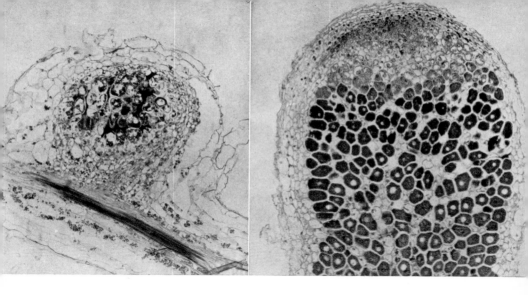

Plate IX. Boron essential for nodule development on leguminous plants: *a* (*top left*), no boron; *b* (*top right*), boron applied. (*Photograph V. Stansfield*); *c* (*bottom*), *Molybdenum* deficiency in New Zealand: Rape: (*right*), on natural soil; (*left*), supplied with 3 oz. per acre sodium molybdate (*W. R. Lobb, N.Z. Dept. of Agriculture*)

Plate X.

added in the manure; over the succeeding 80 years it was 6 per cent only: all the rest had been dissipated into the atmosphere as carbon dioxide, water vapour and nitrogen gas. (Table 15; see also p. 137).

TABLE 15

Loss of organic matter from farm yard manure added to the soil; tons per acre of dry organic matter:

	ADDED IN MANURE	REMAINING IN SOIL, EXCESS OVER UNMANURED[1]	LOSS	PER CENT OF QUANTITY ADDED
After 22 years ..	61	16	45	74
Change in last 80 years	225	9	216	96
After 102 years ..	286	25	261	91

So far as is known plant residues such as are contained in farmyard manure, composts or green manures are the only organic materials that directly increase the amount of organic matter in the soil and even then the process appears to us wasteful. The wastefulness, however, is more apparent than real; the organic matter recorded as lost has gone to support the population of living organisms which, as shown in Chap. 4, is much larger on the manured than on the unmanured soil.

It has been supposed that organic manures, and particularly composts, have some special effects on plant growth which cannot be produced by inorganic fertilizers. The underlying idea is that a nutritive element coming direct from plant residues has some vital quality not possessed by the same element in an inorganic form: a relic of the old vitalistic theory which has long been given up by biologists. There is no evidence whatsoever for this supposition, and no reason at all to think that composts or any other organic manures have any special value apart from physical effects they may exert in the soil and the nutritive elements they contain: usually these are considerably less effective than those in simple inorganic form because

[1] Based on nitrogen figures of p. 26, assuming the soil organic matter contains 5 per cent of nitrogen (p. 23). The weights of fine earth in the top 9 inches were, in tons per acre: Unmanured, 1,160; Manured after 22 years, 1,110; after 50 years, 1,040. No later figures are available.

Plate X. Trace elements essential for plant growth: Pot 1, no copper; Pots 2, 3 and 4 received 6, 12 and 24 parts per million respectively. Pot 5, no manganese; Pot 6, no zinc; Pot 7, no copper; Pot 8, all three supplied. (*C. S. Piper, Waite Institute, Adelaide*)

they all have to be reduced to simple ions before the plant can take them up. No problem in Nature is ever completely solved, however, and some new and hitherto unsuspected property of organic manures may yet be discovered: all one can say is that there is no sign of this at present. Comparisons of compost with fertilizers at Rothamsted and at Wye[1] have shown that fertilizers gave better results.

Like other chemical aids to plant production and plant protection, however, fertilizers can do harm if wrongly used. Knowledge and intelligence are needed to achieve the best results with them. Organic manures make less demand on either, and may therefore be safer, but they are more costly and less effective; they are therefore much less used by large growers than formerly.

Various organic materials have been used for manuring special crops such as hops, fruit, and high quality vegetables; some gardeners still prefer them: rose growers swear by dried blood. Their chief value is as a source of nitrogen and on this they should be judged; for reasons given on p. 68 anything with less than 4 per cent of nitrogen should be used only after careful tests have shown that it is really likely to be effective. Some of the commoner substances are listed in Table 16.

TABLE 16

Composition of certain organic manures: per cent.

	EFFECTIVE NITROGEN	PHOSPHORIC ACID (P_2O_5)	MORE RESTRICTED VALUE.	NITROGEN
Dried blood	12 – 15	—		
Hoof and Horn ..	12 – 14·5	—	Sewage sludge	
Wool waste (Shoddy) ..	3 – 17	—	(Lagoon)	0·5 – 1
Fish guano	7 – 9	3 – 8	Town waste	0·5 or less
Activated sewage sludge up to	6	about 4		
Rape meal, about ..	4	5		
Malt culms, about ..	5	2		
Meat and bone meal ..	5 – 6	13·5 – 16		

[1] *Rothamsted Annual Rpt.* 1954, p. 154: *Empire Jl. Expt. Agric.* (1955) 23, 225 (Wye results). After some years without addition of organic matter the soil at Wye showed signs of falling behind in productiveness, but this only proves the importance of organic matter in conserving the structure of the soil, which is not disputed.

Most of the effective materials have been used for many years and are described in detail by the early writers. Plot in his History of Oxfordshire (1677) speaks highly of "old rags" (presumably woollen) worn by men and women and "well sated with *urinous salts* contracted from the *sweat* and continued *perspirations* attending their bodies". The older farmers preferred woollen rags from poor people because of their higher content of "urinous salts"; but the farm workers objected to chopping them up and handling them because of the danger of catching smallpox. The modern version is shoddy waste from the Yorkshire mills, a mixture of cotton and wool the value of which depends entirely on the percentage of wool as the cotton contains no nitrogen. It is still much used by hop growers.

No method of enriching the soil in organic matter is as effective as leaving it under a mixture of grass and clovers or other leguminous plants for some years. This is widely practised by good orchardists in the south and east of England: the grass is frequently mown and the cuttings left on the surface so that the plant nutrients they contain may return to the soil. (Plate XIIIa, p. 198). It has been shown earlier that leaving the land under grass raises its content of organic matter to a point beyond which it cannot further increase because of the concurrent increases in numbers and activities of the soil population. Cultivation curtails the supply of plant residues and increases that of oxygen: decomposition is speeded up and the amount of organic matter in the soil falls. This reduces the numbers of the soil population and decomposition consequently slows down: ultimately a lower limit is reached below which the organic matter content of the soil does not fall. The limits vary with rainfall, temperature and other conditions. Some of the values obtained for them are:—

	PER CENT OF ORGANIC MATTER IN DRY SOIL		
	SANDY SOILS	LOAMS	CHALK SOILS
Upper limit	4	5	8
Lower limit	0·5	1·8	2·5

Additions of plant residues can keep the soil organic matter at any desired level between these limits, but they must be regularly maintained. They cannot of themselves usually raise it above the upper limit; for this more drastic changes in the soil conditions would be necessary. Usually the limit is higher in moist than in dry conditions because moisture promotes plant growth and consequently greater additions of organic matter to the soil; it is higher also in cool than in hot conditions because the activities of the soil population are less: the effect of temperature, however, appears to be less than that of rainfall. The decomposition proceeds in such a way that whatever the level ultimately attained the ratio of carbon to nitrogen in the organic matter is somewhere about 10; it follows that an increase of one part of nitrogen in the soil cannot be effected without at the same time adding ten parts of carbon. This explains why the inorganic nitrogenous fertilizers nitrate of soda and sulphate of ammonia have added nothing to the nitrogen content of the soil except a small quantity attributable to extra root growth caused by the fertilizer, even though they have been applied annually for many years in relatively large dressings to some of the Rothamsted plots. Conversely the soil does not lose nitrogen unless at the same time ten times the quantity of carbon is lost. This is another example of the buffering of the soil against change.

Plants can be grown perfectly satisfactorily without any organic matter and indeed without any soil. This has been done for something like a century in scientific laboratories: water culture is a standard operation giving full and perfect development; for some years now it has been practised commercially. The first large scale trials made in California in 1936 by F. Gericke created a sensation in the popular Press: "hydroponics" as the method was called, was hailed as a great discovery and some highly exaggerated claims were made for it. Critical tests led to disillusionment, but properly conducted investigations, among others those by W. G. Templeman at Jealott's Hill and more extensively by R. H. Stoughton at Reading, showed both its possibilities and its limitations. Gericke grew his plants in a tank containing a nutrient solution supporting them on a wire net. This proved difficult in practice because oxygen, which is essential for the plant roots, is only slightly soluble in water and is therefore not obtainable in adequate quantities; while carbon dioxide which is evolved by the root and is poisonous, dissolves much more freely and can therefore exert its full asphyxiating effect.

These difficulties are overcome by using a solid medium, rock chippings, gravel, sand, or broken clinker, and flooding it periodically with the nutrient solution which is then drained away. The flooding may be either from above or below; the former procedure involves less capital outlay as it can be done by a hose; the latter can be worked automatically. The method has some advantages over ordinary soil culture. It is much more economical of plant foods as there are none of the interactions with soil constituents or assimilation by soil organisms that reduce the efficiency of the fertilizers applied. It is more economical of labour as there is no winter digging of the soil and no surface cultivation. There is also less likelihood of root diseases, and the small risk can be eliminated by occasional sterilisation with formaldehyde. But the installations are costly, and although the yields are high they do not usually sufficiently surpass those of good soil management to justify the cost for that reason only. The limiting factor in good English glasshouse practice is light, and the method does nothing to overcome this. Where, however, economic factors are not the chief determinants the method has proved useful. It was adopted during the 1939-45 war to provide the United States forces in some of the Pacific Islands with fresh vegetables and salads, and also after the war during the occupation of Japan, when locally grown produce was forbidden to the troops because of the risk of cholera. Installations set up on some of the islands including Okinawa and also on the mainland obviated the necessity of flying fresh produce from California or Honolulu. Fears that the food value of the produce might not be equal to that grown in soil receiving organic manure have proved unfounded Samples of tomatoes grown at Reading showed no diminution of ascorbic acid or Vitamin A, two of the most valuable nutritive constituents of the tomato, and there was a gain of sugar and a decrease of acidity.

The very considerable saving of labour as compared with the old method of cultivation is a great advantage in raising costly glass house crops. Carnations are grown commercially by this method; the perfect control and freedom from soil pests and diseases ensures the production of spotless blooms for the early market. (Plate XXI*b*, p. 220).

The method is much studied at the Experiment Station for Floriculture at Aalsmeer, Netherlands.

CHAPTER 7

MAN'S CONTROL OF THE SOIL:
II. ENVIRONMENTAL FACTORS

DURING THE 19th Century British agricultural scientists concerned themselves mainly with the problem of plant food: the standard of farming until the last decade was so high that the other requirements of the growing plant were usually met. In America, however, other factors especially water supply to the plant were more important, and in studying the problems involved the United States scientists were led to develop a new branch of science, Soil Physics, pioneering work on which had already been begun in Germany. The subject has greatly expanded and it is now much investigated in all advanced countries: at Rothamsted first by B. A. Keen, later by R. K. Schofield and their colleagues and successors. Along with these investigations have been others on plant physiology, and the result is a vast accumulation of knowledge about the requirements of the growing plant, and how the soil can be made to satisfy them.

For full growth the plant requires from the soil (1) an adequate supply of plant food, (2) a continuous and sufficient supply of water and of air, (3) suitable temperature, (4) ample root room, (5) absence of harmful substances, (6) freedom from competition with other plants.

Good farmers and gardeners have arrived empirically at methods of soil management which did in fact satisfy all these requirements to a greater or less extent. Now that the rapidly growing world population necessitates still higher output from the soil it has become imperative to intensify these methods and to devise new ones, which must be based on a clear understanding of the objects that have to be achieved.

The continuous supply of water needed by the plant is taken up by the roots from the pore spaces of the soil, and as replenishment by rain is only intermittent adequate storage within the reach of the root is essential. Air is equally necessary; the total pore space has to be large enough to allow free access of atmospheric air in addition to holding a sufficient store of water. This cannot be attained if the soil particles remain separate and distinct, for then they would settle down to a condition of closest packing with a minimum pore space which, if it was to hold enough water, would not have space left for sufficient air, and in the case of a heavy clay would form a sticky mass when wet and a hard apparently solid block when dry, impenetrable by the roots and incapable of supplying air in sufficient quantity.

The effect of packing on the pore space of a sandy soil is strikingly shown when one is walking on moist sand at the sea shore as the waves are retreating. The sand is left in a state of closest packing; its pore space is the smallest possible and any disturbance can only increase it. Stepping on the sand therefore does this; and it has the unexpected effect that water is sucked in from the surrounding sand leaving a drier fringe around the shoe. When the foot is lifted the sand quickly reverts to the state of closest packing and minimum pore space; it can no longer hold the water it has sucked in but extrudes it forming a little pool. If however it were possible to keep the sand grains in the new position into which they had been forced then of course the pore space would be permanently increased, the reserve of water would be larger, water could more easily move into the subsoil, air could enter and carbon dioxide diffuse out more easily. If the clay particles could be separated the solid impenetrable block would become more permeable alike to air, water and plant roots. This separation of particles is in fact more or less achieved by the various cultivation devices evolved empirically by practical men.

In any attempt to improve on a good empirical process it is first of all necessary to find out exactly what it has accomplished and why. Frequently it happens that the practical man's methods are right so far as they go, but his reasons are wrong: only properly conducted scientific studies can discover why the methods have worked.

Examination of good soil tilths satisfactory to the gardener and the farmer shows that the soil has been built up into crumbs of about 0·5 to 5 mm. diameter, each like a minute sponge permeated with fine capillary passages, and in between the crumbs are considerably larger spaces making channels down which water can drain away and air can enter. Under these circumstances seeds can germinate well, the seedlings prosper and a good system of white feeding roots can develop.

If time were no object the building up of such crumbs would present no practical problem because they are formed naturally under grass and clover sward. How exactly this is brought about is not yet known, but it probably results from the action of earthworms and of the soil micro-organisms on the organic matter sloughed off from the plant roots. Large strong roots appear to be more effective than fine ones.

The important question is, however, how long does the natural process take? This has been studied by A. J. Low of the Jealott's Hill Station[1] who finds that on coarse sandy soils a period of 5 to 10 years may be sufficient to give a structure as good as under very old grass, but on clay soils much longer time is needed, even up to 50 years. On poorly drained soils the process is retarded because root development is weakened, and heavy grazing causes retardation by compacting the soil too much. Table 17 shows how slow the process can be: it gives the percentage of soil crumbs stable in water (the mark of a good crumb) present after various periods under grass.

TABLE 17

Percentage of water stable soil aggregates larger than 3 mm. diam. in soils under grass for different periods:

	PERCENTAGE OF AGGREGATES	ORGANIC MATTER: PER CENT	SOIL MOISTURE WHEN FULLY DRAINED: PER CENT BY WEIGHT
Old pasture, 100 years	49	10·4	35·6
Old arable, then grass for 28 years ..	27	5·7	30·8
Old arable, then grass for 6 years ..	18	4·0	24·6

For most people this natural process is too slow and the crumbs have to be formed by cultivation. There are two distinct problems: the crumbs have to be made, and their stability has to be ensured. The building up depends on the arrangement of the soil particles, the stability on their surfaces and the forces holding them together. The building up requires some cementing material and of this there are two kinds: mineral and organic. The most important mineral cement in our climate is clay, especially of micaceous type with an appreciable power of holding bases: the kaolinitic clays are less effective. In the tropics hydrated ferric oxide is frequently the chief agent. The organic cementing material includes a number of products of micro-organic

[1] Jl. Soil Sci. (1955) 6, 179–199.

action, polysaccharides and others; also the bodies of some of the organisms themselves, living or dead.

Two methods have long been practised on light land for building up sand particles into crumbs. One is to add clay or the mixture of clay and calcium carbonate called marl:[1] the latter having the advantage on sandy soils when as often happens they are acid. Much of the light land of the eastern counties, both sands and fen, too light for cultivation and therefore left as barren heaths, was reclaimed in this way in the latter part of the 18th Century and through much of the 19th. Large dressings of clay of the order of 100 tons or more per acre were needed; fortunately clay underlies much of the fen soil so that transport problems did not arise. The improvement lasts for many years.

The other method makes use of the much weaker cementing material, organic matter, and it has to be combined with consolidation. It consists in folding sheep on the land. This method also goes back to very early times but it was greatly developed in the late 18th and the 19th Centuries after the introduction of turnips and other fodder plants as farm crops for animals, and the drilling of seeds in rows which allowed of cultivation between them. A crop was grown for sheep to eat; a number of them were penned on to a small area until they had completely cleared it; they were then moved a little further on and this continued till the whole field had been cleared. Their droppings manured the land and their feet consolidated it: the combination was almost perfect. An adult sheep may weigh anything from about 200 to 300 lb. according to its breed, and a lamb 9 months old may weigh 150 to 200 lb. This weight is supported on four small feet so that the pressure each exerts per square inch may be some 40 to 70 lb. As the sheep are rather closely packed and move about a good deal there is considerable treading of the soil in both wet and dry conditions: and the resulting consolidation is of a distinctive and effective type that no implement can entirely reproduce. So great was the benefit to the succeeding crop and to the condition of the soil that farmers regularly spoke of the "golden hoof" as the best amendment for light soils. The high cost of labour, and the difficulty of getting shepherds, have combined to make folding impracticable so that it is now much less practised than it was, but no equally effective method of improving a light soil has yet been found.

A third method was used on heavy soils: it was to add large

[1] The marls of the geologist, e.g. Keuper Marl, are not necessarily calcareous.

quantities of chalk—calcium carbonate—which makes the clay much less sticky, more readily permeated by fine cracks and broken down into crumbs. Water more quickly drains away and the land is more easily cultivated. The method was brought into England before the Romans came; whether it was introduced by the Iron Age Celtic peoples or by still earlier invaders is not known. It was new to Pliny who thus describes it:—

"The peoples of Britain and Gaul have discovered another method for nourishing the land. There is something they call marl (*marga*). It contains a more condensed richness, a sort of fatness of the land . . . Marl is twofold in nature: hard or fatty; these can be distinguished by the touch. In like manner it has a two fold use; some kinds are used for grain crops (*fruges*) only, others for herbage (*pabulum*) . . . of the fatty marls the chief are the white varieties. There are several of these . . . (one) is the silvery chalk. It is sought for deep in the ground, wells being frequently sunk 100 ft. deep with the mouth narrow and the shaft widening out as in the mines. This is the kind most used in Britain. It lasts for 80 years and there is no instance of anyone who has put it on twice in his lifetime". (Pliny Secundus, Caius, *Naturalis historia*, Book 17. 6.)

The method long continued in use on the heavy soils overlying the chalk in the eastern and southern counties of England. Right up to the 19th Century the owner of such land on coming into his inheritance would chalk it heavily enough to last his lifetime. When the work was done the wells were as far as possible filled in leaving the dells so characteristic of these districts. Fields at Rothamsted chalked probably a century and a half ago still retain some of the calcium carbonate then added: the natural soil has none, and is acid, sticky and difficult to work, while the chalked soil in places contains as much as 2 per cent of calcium carbonate and is more friable and tractable. The method remained in use right up to the time of the First World War: it was a traditional craft carried out by gangs of itinerant workmen who were very skilful at finding the best points for sinking the wells—for the underlying chalk has a piped, not a level surface. In 1915 I engaged a gang to chalk one of the Rothamsted fields; they did it exactly as Pliny had described.

The action of the calcium carbonate is not fully understood. There is apparently some flocculation or increase in stability of the crumbs: after liming a Cambridge clay soil the drainage water was clearer and more free from silt and clay particles than before. But this does not appear to be the whole story, for some very difficult clays, such as the Gault, contain quantities of calcium carbonate, and some of the chalk soils are apt to dry to hard steely lumps which require considerable skill to break down properly.

The methods now used for developing soil structure are modifications of the old cultivation devices. The implements have been improved out of all recognition, and while it might be argued that craftsmanship is not what it was there is little doubt that greater control is now possible than in the old days. The engineering problems have been studied at the Agricultural Engineering Institute at Silsoe in Bedfordshire, and the scientific problems at Rothamsted and at Oxford by E. W. Russell.

Light soils are given as much organic manure as possible; it decomposes very quickly and the soils are described as "hungry"; but while it lasts it keeps the soil in a good crumb condition. In E. W. Russell's view the effective agent is not a sticky product of decomposition as has often been assumed, but the finely divided fibrous material which prevents the sand particles falling into a condition of closest packing. The cementing is effected by clusters of fine particles, presumably clay which can be seen in the crumbs between the sand grains; the surfaces of the grains, however, appeared to be clean.

Heavy soils are more difficult to manipulate. The soil must first be opened up by either the spade or the plough, preferably in autumn or early winter, and left in coarse lumps to the action of the weather. The medieval ploughs turned up large lumps that had to be broken by mallets as shown in the beautiful 15th Century Luttrell Psalter (Plate XII, p. 179); the modern ploughs break the furrow slice into smaller clods. For making the crumbs the operative agents are climatic, and the purpose of cultivation is to put the soil into such condition that they can act most effectively.

Frost is the best of all agents especially if it comes slowly; the soil should be dug or ploughed in late autumn or early winter and left in rough state so as to expose the greatest practicable surface to its action. As the water in the clods cools it reaches its maximum density at 39°F. and on still further cooling it expands, particularly when the freezing point is reached. The pressure exerted within the clods by this expansion reduces the pore space and so forces out some of the water held therein. Alternate freezing and thawing, particularly if slow, thus somewhat dries the soil, which conduces to stable crumb formation. Even when there is no frost, however, the alternate wetting and drying effected by the rain and the wind prompts the formation of stable crumbs: the resulting swellings and contractions cause cracks which weaken the clods and if now they are struck by a hoe or a harrow while just sufficiently moist they fall into smaller fragments which can be further worked down.

The structure of heavy soils is improved by addition of organic matter like farmyard manure, compost, or rape cake, but large quantities are needed to produce measurable effects. On the Broadbalk wheatfield the plot that has received 14 tons of farmyard manure each year since 1843 now contains 55 per cent of water-stable crumbs greater than 0·5 mm. diameter, while the adjacent plot left without manure over the same period contains 28 per cent only. For a corresponding pair of plots on the rather heavier Barnfield soil cropped with mangolds the figures are 70 per cent on the manured and 54 per cent on the unmanured plots. The occasional small dressings usual in farm practice produce nothing like these results, though impressive dressings of the order of 100 tons per acre such as are sometimes given in commercial horticulture may prove very effective.

Only the easily decomposable organic matter seems to act well: apparently it coats the clay particles, preventing them from sticking together firmly to make a hard clod, though leaving them with sufficient cohesion to build up crumbs. Although commonly regarded as a cement, the organic matter functions rather as a separating agent both in coarse sandy soils and in heavy clay soils: in the former case the operative material appears to be the fibre, in the latter the easily decomposable material. It is possible that bacteria and fungi in decomposing the organic matter assist in forming soil aggregates by attaching themselves to the smallest particles forming clusters: the effect may not long survive the death of the organism, however. The loams which come between the sands and the clays are more readily worked without organic manure than either, and they provide many examples of successful farming with no livestock. The effect of organic matter on a clay soil is only transient; the active material is produced by the soil organisms but it is also easily decomposed by them.

The soil conditioners referred to on p. 32 have the advantage of resisting attack by micro-organisms, and in consequence are needed in small amounts only. It is claimed that a quantity of IBMA or VAMA equal to 0.05 per cent of the weight of soil produces the same level of soil aggregation as 2 to 5 per cent of organic residues—40 to 100 times the weight of the conditioner. The quantity recommended is 30 lb. per acre if the whole soil is covered, 2 to 5 lb. per acre if only strips are treated in which seeds are sown, and of course less if only isolated patches are treated. The effect is claimed to persist for three years or more.

Soil structure once built up has to be preserved. It may be destroyed by heavy rain, or by working the soil when it is either too wet or too dry. The fine particles break away from the crumbs, and run together forming a glazed surface which must be broken as it impedes the movement of air and water into the soil. A mulch[1] affords good protection against heavy rain; it also keeps the soil moist, so facilitating the activities of the soil organisms; in an Oregon experiment a soil covered for five years with a mulch contained at the end of the period 34 per cent of water stable aggregates larger than 1 mm. while in adjacent soil carrying weeds and grasses the percentage was 17, and in soil kept bare of all vegetation it was only 3. Lawn mowings scattered over a bed of shrubs or herbaceous plants not only prevent the formation of the glazed surface but also help to keep down weeds, so saving hoeing. As soil structure is most important in the upper layer of the soil it is there that control is most needed.

The investigations by E. W. Russell to find out how cultivation improves the soil have led to some unexpected results. Barley has always been regarded as requiring a good seed bed and a fine tilth, yet the samples submitted by farmers to the Malting Barley Conferences at Rothamsted showed that equally good yields and equally good quality were obtainable whether the seed bed had been judged good or bad. Cultivations broke the surface of the soil and reduced the lumps, both of which processes were necessary for the establishment of plants, but they had little effect on the pores between 4 and 50 μ equivalent diameter which supply water to the plant, or on the larger pores along which air passes to the roots and carbon dioxide escapes. The fineness of the tilth seemed to be of less consequence than had been supposed;

[1] A mulch is a covering, usually of organic matter: leaves, lawn mowings, farmyard manure, and the like.

what was important, however, was the complete destruction of weeds, which were very harmful to the young plants.

It was for long thought that hoeing the soil to cover it with a fine mulch would conserve its moisture but this does not seem to be the case: H. C. Pereira after examining all the evidence could find no clear proof that it is so.

Increasing the depth of cultivation is not by itself always advantageous. In E. W. Russell's experiments deep ploughing was of little benefit to potatoes on heavy soil at low levels of manuring, though it was beneficial at higher levels. The effect was reversed on light soils; deep ploughing was more beneficial where yields were low than where they were high. It was not, however, beneficial on undrained soil as it increased the volume of water held by the soil so excluding air and keeping down the temperature. Subsoiling acted much like deep ploughing but was not so effective for potatoes in heavy soils though it sometimes acted well on light soils, presumably by breaking up the compacted layer made by the tractor wheels on the furrow bottom. For sugar beet, however, deep cultivation was beneficial, and subsoiling seemed even better than deep ploughing. A pan or plough sole must of course be broken.

Deep cultivation by trenching in the garden is justifiable in two sets of conditions: where a hard layer or pan is impeding drainage, and where the soil is to be made more suitable for deep rooting plants. In the former case the whole drainage system has to be examined to ensure the final removal of the water; in the latter case plant residues, compost or farmyard manure must be worked in to the lower spit to provide the basis for an active soil population. Experiments by S. U. Pickering and the writer showed that trenching by itself caused no improvement in the growth of fruit trees either on a sandy soil or a heavy clay.

In a considerable number of experiments made by E. W. Russell on many farms in different parts of the country the one consistent benefit from cultivation was the destruction of weeds. Few gardeners or farmers realise the harm done by weeds, particularly young weeds which are apt to be left alone because they are so small. Modern selective weed killers, especially those that kill the weeds before they emerge above the surface of the soil, may save a considerable amount of cultivation.

CONTROL OF THE WATER SUPPLY

The water supply in the soil comes ultimately from the rain which is at present beyond control. Some is stored in the upper layers of soil, some drains away and some evaporates: these three processes can to some extent be controlled. As explained in Chapter IV the storage is in the pore spaces; the maximum amount that a growing plant can extract is the difference between the quantity held when a fully wet soil has just ceased to lose water by drainage and the amount remaining when wilting begins. Expressed in inches of water[1] per foot depth of soil these quantities are:—

	COARSE SANDY SOILS	LOAMS SANDY	LOAMS SILTY	CLAYS
Uniform depth ..	0·50 – 0·75	1·25 – 1·50	1·75 – 2·25	1·8 – 2·0
More compact subsoil	0·75 – 1·0	1·25 – 1·75	2·0 – 2·25	

The sandy soils are better than the figures suggest because plant roots grow so freely in them; their growth is much more restricted in a clay soil. The available reserve of water in the soil when summer begins in April cannot be reckoned at more than about three or four inches; if the summer is dry plants may suffer from lack of water.

Irrigation is as old as civilisation and was indeed a powerful agent in its development. A great amount of empirical knowledge has been acquired about it, and the Egyptian fellahin in the Nile valley have an uncanny facility for levelling their fields before running in the water so that no one patch shall have more or less water than any other.

Much of the modern scientific work has been done in Utah by the Mormons and in California. As water is likely to be scarce in any region where irrigation is necessary it is essential to know how much should be given to ensure full growth. F. J. Veihmeyer in California has worked out a method by which this can be done. He determines experimentally the water content when wilting begins, assumes that all water in excess of this amount is of equal value to the plant, and as excess would be wasteful he recommends a dose of water that ensures an adequate but not too large a margin above the amount present at the wilting point.

Irrigation in Great Britain is essentially a modern development

[1] One inch of water = approx. 100 tons per acre: more accurately 101.3 tons, or 22,622 gallons or 3,630 cubic feet per acre or 4½ gallons per square yard.

though the watering of meadows has long been practised in some of the midland and southern counties. The scientific work has been done chiefly at Rothamsted. The approach differs from that of Veihmeyer in that it has been from the other end of the scale, the point of maximum water content when drainage has just ceased. As the need is for high crop yields and the quantities of water required are small compared with those of the great irrigation systems of the world the aim is to keep the plant as near as possible to a state of maximum transpiration because many plants make their maximum growth when the amount of water in the soil is near the field capacity. The amount by which the water content of the soil falls below this maximum is called the water deficit, and the purpose of irrigation is to ensure that the deficit never becomes so great that transpiration is seriously checked.

The map of Fig. 10a, p. 180 shows the number of inches of water plants would transpire if maximum transpiration continued during the six summer months April to September. The adjoining map shows the average rainfall in inches during the same period: much of the southern and especially of the south eastern part of England obviously does not receive enough rain to furnish the necessary water. Something can of course be drawn from the soil: presumably it is filled to capacity at the beginning of April and it need not be in this state at the end of September: indeed it is better not, for the production of seed, the ripening of wood, and the healthy dying down of herbaceous foliage with translocation of food reserves to the roots or tubers goes better in dry than in wet conditions. But it is not safe to assume that more than about 3 or 4 inches of water can come from the soil, whatever of the remaining deficit the rain has not supplied must be provided by irrigation.

Before discussing how this is best done it is interesting to know how the transpiration map was prepared. It is not the result of direct measurements on plants but is deduced from meteorological data by the method worked out by H. L. Penman of Rothamsted. The starting point was the amount of evaporation from a free water surface. This is measured at various meteorological stations up and down the country, and Penman showed that it could also be calculated from the long-term average data for hours of sunshine, temperature, and wind velocities. The amount of evaporation from a free water surface was found to be closely related to the amount of transpiration from an equal area covered with vegetation since transpiration depends more

Plate XI. Pot culture house for detailed studies of plant nutrition. Rothamsted.

V. *Stansfield*

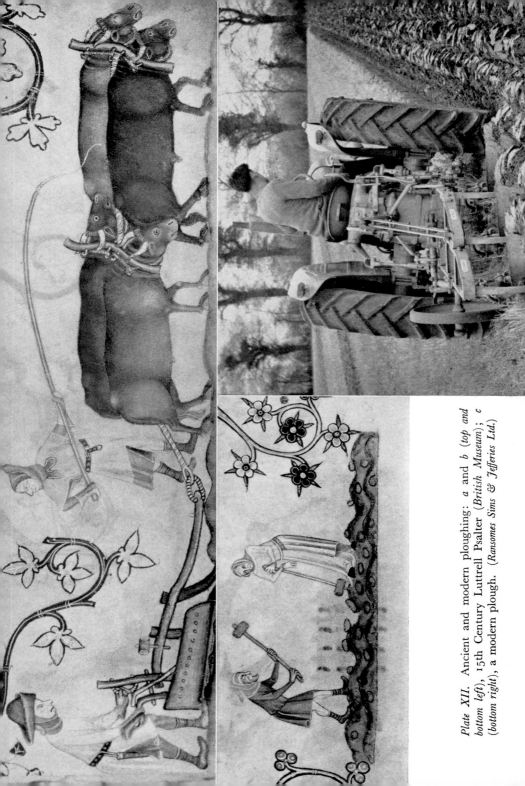

Plate XII. Ancient and modern ploughing: *a* and *b* (*top and bottom left*), 15th Century Luttrell Psalter (*British Museum*); *c* (*bottom right*), a modern plough. (*Ransomes Sims & Jefferies Ltd.*)

on meteorological conditions than on the plants: the ratio of trans-piration to evaporation was 0·6 in winter (November to February), 0·8 in summer (May to August), and 0·7 in spring and autumn (March—April, and September—October). Penman applied these ratios to the data for evaporation from a water surface, made also certain topographical corrections, and so constructed the map.[1]

In certain districts the map could be checked by a totally different method. Some of the Catchment Area Boards have numerous regional data for rainfall, run-off of water, and change in amount of water stored in the soil: the difference between the sum of the two latter and the rainfall equals the amount of evaporation. Where valid comparisons were possible the agreement was satisfactory.

Comparison of the two maps shows that even on the average the summer rainfall in the southern part of England, and especially in the southeast, is 6 to 9 inches below the potential transpiration: in a dry season the deficit may rise to 12 inches or more. Transpiration is highest in June and July when it is about 3½ or 4 inches a month, a water requirement of one inch in ten days: in the dry summer of 1955 the average rainfall over England and Wales during the whole of July was only 1·1 inches and the result was that potatoes rose to over 6d. per lb. On the sandy soil at Woburn the addition of 6·3 ins. of water by irrigation had increased the yield of main crop potatoes from 11 tons to 20 per acre. Had irrigation of the potato acreage been general there would have been a glut instead of a scarcity.

Growers of expensive crops have long adopted methods for supplying the necessary water. In the Vale of Evesham, famous for its production of fruit and vegetables, many growers have permanent provision for irrigation; some of those on the Avon have electrically driven pumps, reservoirs and underground pipes: installations which may have cost some £20,000; others further away have sunk wells to supply the water needed. The water is sprinkled on to the crop from spray lines or automatically moving spray guns; at a rate of ¼ or ½ inch per hour the water can penetrate the soil properly without damaging its structure.[2] Commercial growers commonly judge empirically how much

---

[1] The method of calculation is given in *Q. Jl. Roy. Met. Soc.* (1950) 76, 372–383, and Tables showing monthly transpiration values for different parts of the country are given in *The Calculation of Irrigation need*, Tech. Bull. No. 4. Min. Agric. 1954, H.M.S.O. A small change has since been made in the method of cultivation which reduces the transpiration estimates by about 5 per cent: roughly 1 inch per annum.
[2] Practical details are given by C. S. Wright, *Jl. Farmers' Club*, 1956, Pt. 3. 26–42.

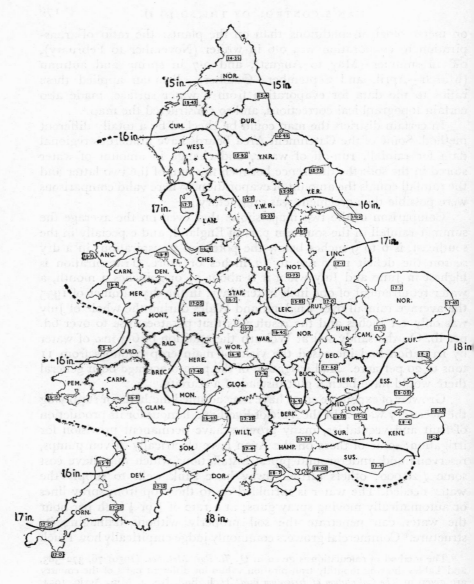

FIG. 10A

Average summer potential transpiration in inches, April–September, for the period
1891–1915.

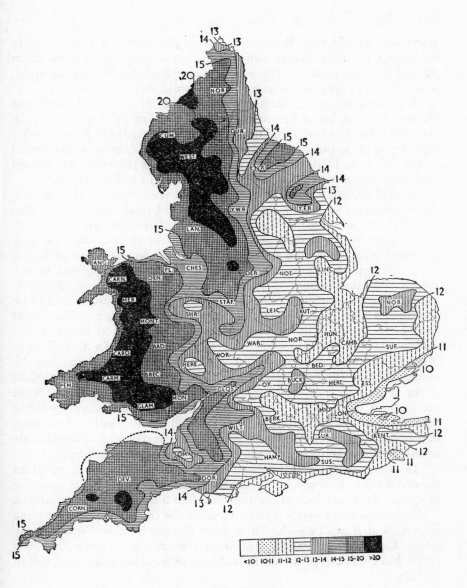

FIG. 10B

Average summer rainfall in inches, April–September, 1891–1915.

(By permission of H.M.S.O.)

water they should give; Penman has shown, however, that the quantity
calculated from the water deficit allowing something to come from
the soil is a safer guide as it avoids the waste involved in over-watering,
and the loss of crop consequent on an insufficient supply.[1]

Watering by spray simulates the action of rain and need not
damage the soil, but it involves a good deal of loss by evaporation. In
the Netherlands where irrigation is much used, resulting in amazingly
high levels of outputs, as much as two-thirds of the spray water is said
to be thus lost. Underground or subirrigation is therefore much used
and some ingenious methods have been devised for this purpose.

In arid regions, especially those in tropical or subtropical countries,
great irrigation schemes have been set up, and more will be needed as
the world population increases and more food is required. Only a few
of the great rivers of the world are suitable for the purpose: the best
are those rising from the Himalaya mountains in the north of India,
they account for about one-seventh of the irrigated regions of the
world. But the amount of water needed in these arid conditions is
very great. H. Addison[2] estimated that 4,000 tons of water would be
vapourised during the growth of a ton of maize in a subtropical
climate: a square mile of land so cropped would feed between 600
and 700 people at a cost of $1\frac{1}{4}$ million tons of water. In arid climates
only a fraction of the rain falling in the catchment area reaches the
rivers, the rest simply evaporates: in the Murray river system less than
3 per cent is so recovered; this compares with about a 30 per cent
recovery in the Thames basin. Ultimately the water supply is likely
to limit the world's production of food from the soil unless some
economically feasible means of desalting sea water on a colossal scale
could be devised: all known methods are at present too costly.

In many arid regions subsoil water is available but it contains
dissolved salts. Few species of plants tolerate any appreciable quantity
of these; plants are indeed less tolerant than human beings or animals.

The evaporated water is not all lost: in the cool of the evening some
is precipitated as dew—the "mist from the earth" that "watered the
whole face of the ground". Modern research has shown that this
really has come out of the earth and not, as used to be thought, out of
the air.

[1] He gives an example in *J. Agric. Sci.* (1952) 42, 286–292 (irrigation of sugar beet).
[2] *Land, water and food*, p. 25, (1955) London (Chapman and Hall).

CONTROL OF AIR SUPPLY IN THE SOIL

The movement of air in and out of the soil, the functions it performs there and the part it plays in the assimilation of nutrients from the soil by the plant roots are described in Chapter 3. The air supply is contained in the larger pore spaces of the soil but so also is water which is equally necessary to the plant: a proper balance has to be maintained between them. For a light or medium soil removal of water brings in an equal volume of air; for a heavy soil the volume is rather less because as water leaves the fine pores by evaporation the clay shrinks and the pores to that extent collapse. Fortunately it happens that the needs of most plants are satisfied by the water held in a fully drained soil—the field capacity of the preceding section—and this allows ample space for the air. The two methods by which the air supply in the soil can be increased are the increase in the pore space and the provision for unimpeded drainage.

The increase in the pore space is effected by loosening the soil and also by the building up of soil crumbs which as already shown is largely a matter of cultivation and the provision of organic matter. But the loosening must not go too far. Plant roots can develop properly only in a fairly compact soil: indeed when the soil is moderately dry fruit trees and shrubs can be rammed in almost as if they were gate posts. Moreover soil moisture readily evaporates if the soil is too loose, particularly in the eastern and south eastern part of England where the rate of evaporation is high; rolling the arable land to compact it after sowing the seed is a regular farm operation, and gardeners know well that the soil must not be too loose—or as the older Kentish men called it, too "hover". There can, however, be no compromise on the matter of drainage and this should be done as effectively as possible. Drainage used to be an empirical craft; it has been put on a scientific basis by investigations in the United States, the Netherlands and Germany; good work also is being done in the Cambridge School of Agriculture.

CONTROL OF SOIL TEMPERATURE

The temperature of the soil depends partly on the amount of heat received from the sun, and partly on what becomes of it when it gets into the soil: the various factors are discussed in Chapter IV. The heat—more strictly the energy—arrives in short waves but only by day;

part of what the soil absorbs is lost as long waves not only by day but by night as well. Short waves can pass through glass, long waves cannot to anything like the same extent: a glass covering is therefore an effective heat trap, resulting in a higher sunset temperature than in the open.[1] So gardeners use cloches, frames, Dutch lights, cold glass houses and various other devices to get the utmost value from the sun's radiation. H. W. Miles at Wye has worked out some sequences of valuable early market garden crops achieved in this way. The higher temperatures and extra growth necessitates the use of much additional water; considerable quantities of farmyard manure or compost—100 tons per acre or more—must be applied to keep the soil in good condition. Mediterranean plants like vines and peaches may require some additional heat; cucumber growers aim at temperatures of about 85°F. by day and not less than 65°F. by night. On hot summer days the temperature in the glasshouse may rise to 110°F. which would certainly injure the plant if it did not cool itself by transpiration. Copious watering is therefore necessary; some growers also shade heavily. A good yield is 120 tons per acre; some of the Lea Valley growers used to get 180 tons per acre but such yields are no longer economic.

Other crops grown commercially under glass for the sake of the extra warmth and protection against wind include roses, carnations, mushrooms and winter lettuce, this last is very popular in Lancashire.

Heating the soil by electricity is practised by growers of seeds and flowers in the Netherlands, also for the forcing of chicory and the raising of tomato and cucumber plants. A bed is dug out 4 to 8 inches deep, strips of galvanised wire meshing are laid down but not allowed to touch, each is then connected with the electric current.

Drainage is the most effective way of increasing the temperature of a wet soil.

ELIMINATION OF COMPETITION BY OTHER PLANTS

It has long been known that plants do not readily tolerate competition unless there happens to be a great abundance of food and water. The old idea was that plants injure each other by excreting from their roots substances harmful to plants of the same kind but not necessarily to those of another kind: the advantage of rotation of crops

[1] See *Weather and the land*, Bull. 165, Min. Agric. pp. 16–17. The night cooling under glass is about the same as in the open, but having started at a higher level the temperature does not fall so low. The conductivity of the glass is important.

was thus explained. Some plants were supposed to be more generally harmful; in John Hunter's edition (1787) of Evelyn's *Terra* there is a description of the Scythian or Tartarian lamb which had "something like four feet, and its body is covered with a kind of down. Travellers report that it will suffer no vegetation to grow within a certain distance of its seat". He gives an illustration of it which is reproduced in Plate XIII*b*.[1] The idea of a toxic excretion from plant roots was later given up but was revived by Spencer Pickering, and his experiments have never been proved faulty. There is evidence that one clover plant may impede the growth of another by suppressing the formation of nodules.[2] Whether or not harmful substances are excreted there is certainly competition for available water and nutrients, and all unwanted plants—called weeds for short—should be ruthlessly eliminated. As already pointed out, this is one of the chief benefits of cultivation.

Chemists have long sought for substances that would destroy weeds without harming desirable plants, and visionary as the aim might appear they have had considerable success, particularly in protecting grass, wheat and other cereals growing in spike-like fashion, while their common weeds have broad leaves and often, like daisies and hawkweed, flatten themselves out against the soil. The earliest weed destroyers were direct plant poisons; they lay much more easily on the broad leaves of the weed than on the narrow upright spikes of the cereal; they killed the weeds, and often did some damage to the crop, but usually this was speedily made good. Solutions of copper sulphate or sulphuric acid were sprayed over a young barley crop infested with charlock, or it was dusted with cyanamide, which had the advantage that after killing the charlock it became converted into a nitrate which served as a valuable plant food.

These older materials still have their uses in spite of their dangerous character: sodium arsenite, very poisonous to human beings and animals, is highly effective for keeping gravel paths clear of weeds; sulphuric acid, horribly destructive, kills weeds that resist all other agents; sodium chlorate, which makes wood, sacking and other such materials highly inflammable, also kills many resistant weeds. Rather large quantities of these materials are needed: some 50 to 120 lb. per acre of sulphuric acid; 100 to 200 lb. of sodium arsenite and 250 to

---

[1] Apparently an addition by Hunter: it is not mentioned in Evelyn's editions of 1676 & 1678. The reproduction faces p. 198.

[2] *Rothamsted Annual Rpt.* 1950 .p. 53.

400 lb. of sodium chlorate; they are dissolved in water and applied as spray. But for general use they are now largely displaced by some remarkable new substances, so potent that only small quantities per acre are needed, and yet so highly selective that little or no harm to the crop need be feared if proper care is taken.

Three groups of these substances are being widely used. The most interesting in many ways are the hormone herbicides, chlorine derivatives of the growth-promoting hormones produced in plants: they are manufactured from benzene derivatives obtained from coal tar. Owing to their complex structure their chemical names are long and so they are always known by their initials. The two best known are 2,4D (2, 4-dichlorophenoxyacetic acid) and its methyl substitution product MCPA (2-methyl-4-chlorophenoxyacetic acid). They are harmless to human beings and animals, they do not stain clothes, and they are not inflammable. They are readily absorbed both by roots and leaves, and then translocated in the plant, and they are so potent that only 4 oz. to 2 lb. per acre are needed to kill susceptible weeds. Unfortunately not all weeds are susceptible. The older samples of MCPA contained as impurity chlorcresols which tainted tomatoes; modern samples are free from this.

Another group of herbicides, the nitro compounds, includes DNC (3,5-dinitro-o-cresol); these stain skin and clothing yellow and are poisonous to men and animals, but they are very potent and kill cleavers, corn marigold, mayweed, speedwell and spurrey which the hormone group cannot do. But neither they nor the hormone group can kill the pestilential Ground Elder (*Aegopodium podagraria*), Couch grass (*Agropyron repens*), Blackbent, wild oat, onion couch (*Arrhenatherum elatius*), sorrel and pignut (*Conopodium majus*). An almost unlimited number of substances of similar chemical pattern to the hormone and nitro-compounds can be prepared and there is always the hope that one hidden in the future will deal with these weeds; meanwhile persistent digging out is the only sure remedy.

Modern spraying implements produce a fine mist that easily drifts and may do considerable damage to a neighbour's plants: already numerous complaints have been made. The destruction of roadside verges by herbicides practised in some places may do unexpected harm by destroying beneficial insects or otherwise disturbing the balance of Nature. Two serious lawn weeds are at present beyond the reach of either group of herbicides: moss, and pearlwort: they still have to be treated with the old fashioned lawn sand, a mixture of 3 parts of sulphate of ammonia with one of calcined ferrous sulphate used at the rate of 1 oz. per square yard on isolated patches, or mixed with 20 parts of sand and given at the rate of 4 oz. per square yard for larger areas. Late spring or early summer is the best time to do this, preferably on a dewy morning in fine warm weather. Addition of the chlorides of mercury is said to make the mixture more potent.[1] The reduction in plant growth resulting from competition is not always harmful, however. The grassing down of orchards not only enriches the soil in nitrogen and organic matter (p. 25) but it affords a safe way of manuring the trees.

If the soil were bare nitrogenous fertilizers might lower the quality of the fruit and potassic fertilizer might induce magnesium deficiency. Applied to grass, however, these things do not happen. The grass grows vigorously but it is repeatedly mown and the cuttings are left on the surface to decompose: the plant nutrients then liberated are absorbed by the tree and the quantity of fruit increases but its quality does not suffer. If the grass were allowed to grow tall it would compete with the tree for water and nitrogen, the water effect being particularly harmful. 10 to 15 mowings may be needed, the decomposed material improves the physical condition of the soil. (Plate XIIIa).

CONTROL OF THE SOIL POPULATION

Control of the soil population has been attempted in two directions: the elimination of particular organisms harmful to plants, and a more general control with the purpose of increasing the production of plant food in the soil. Most of the work has been done on harmful organisms; the methods adopted are described in Chapter 4 under the headings of the individual organisms. They fall into two groups: direct methods of elimination, and indirect methods which aim at tilting the balance of conditions against the organisms that one wishes to suppress.

Direct methods were the first used. The pioneer was a French vigneron, Oberlin, whose vines suffered from root aphids (*Phylloxera*); in 1877 he injected a then well known insecticide, carbon disulphide,

---

[1]Full information on modern chemical methods is given in *Weed Control Handbook* (1956) London, British Weed Control Council.

into the soil to kill them. He succeeded, and indeed did more, for he observed that the grapes responded as if they had received a dressing of manure. Carbon disulphide is, however, unsuitable for practical purposes; it is very volatile and therefore wasteful; it is highly inflammable and the commercial material has an unpleasant smell, besides being poisonous. Many glasshouse growers troubled with parasitic eelworms sterilise their soils by heat, which is effective, but very costly and practicable only for crops of high commercial value. Cresylic acid is much used; chloropicrin is better but it is disagreeable material to handle. Modern chemical agents are regularly used against particular organisms, especially wireworms and leather jackets. Harmful soil fungi are not usually destroyed by agents that kill animals but they are greatly checked by formaldehyde, which, however, does not kill insects though it much reduces nematodes. Some of the complex organo-sulphur compounds are promising but are not yet in general use.

Attempts at more general control of the soil population with a view to increasing the production of plant food were made by E. J. Russell and H. B. Hutchinson. The method used was partial sterilisation by steam or volatile antiseptics: toluene, and later chloropicrin or formaldehyde. The investigations revealed the presence of predators which were keeping down the numbers of the useful organisms: the treatment eliminated them and also increased the production of plant food.[1] The biological effects are complex, and chemical changes are produced in the soil which alter it as a habitat for the organisms: heat liberates ammonia which, until it is nitrified, is harmful to seedlings; on certain soils also it raises the amount of soluble manganese. Sterilising agents are more expensive than fertilizers so that there is no economic advantage in increasing plant food in this way. Control is bound to come, however.

The indirect methods of control consist mainly in altering the ecological conditions so that they become less favourable for the unwanted organisms and more favourable for its predators or competitors. More organic matter may be added to the soil as in the case of the parasitic root fungi described on pages 89-90; or the soil reaction may be changed to become more favourable for the plant and less favourable for the organism (p. 91). The resistance of the plant may be strengthened by improving its nutrition or giving it more space.

[1] Later experiments in forestry nurseries are described in the *Rothamsted Annual Report* 1954, p. 48.

CONTROL OF SOIL REACTION

As explained on p. 50, the reaction of the soil is measured by its pH: the value 7 signifies neutrality; lower values signify acidity and higher values alkalinity. In humid climates such as our own soils tend to be acid because the rain washes out calcium, the great neutralising agent, from the soil, and our common plants grow well at a pH of about 6.5 to 7. Soils in which the values are widely different from these present certain difficulties. It used to be thought that acidity and alkalinity were themselves harmful, i.e. that the plants could not tolerate H or OH ions. This is not so: in water culture plants tolerate a pH range from 4 to 8. The real harm is done in other ways. Acid soils lack calcium and are thus unsuitable for many plants. Acid soils may also contain much soluble aluminium and manganese which many plants cannot tolerate: the aluminium because it accumulates in the roots and presumably impedes the assimilation and translocation of phosphate which may cause phosphate starvation even in the midst of plenty; and manganese because it accumulates in the tissues and interferes with metabolism. On the other hand in alkaline conditions phosphate, iron, manganese, boron and some other trace elements become insoluble and the plant may not be able to get enough and suffers accordingly. Uptake of potassium may also be impeded. These effects are shown in Fig. 11, p. 191: one of their results is that the response to potassic and phosphatic fertilizers is greater on somewhat acid than on neutral soils.

Plants vary in their susceptibility to these various deficiencies and a pH unsuitable on one soil may not be so on another. Sugar beet and potatoes both need ample supplies of calcium, but their reaction to an acid soil would depend on whether soluble manganese or soluble aluminium predominated: if the former, potatoes would suffer more than sugar beet; if the latter, sugar beet would suffer more than potatoes. Barley and the brassica family have less need of calcium, but towards soluble manganese and aluminium barley behaves like sugar beet and the brassicas like potatoes.

Alkalinity troubles are at a minimum in our climate on heavy soils, among other reasons because the carbon dioxide evolved from the roots and by the micro-organisms surrounding them reduces the pH; they are more marked on light soils because the carbon dioxide more readily diffuses out. The group of plants called by gardeners "acid loving" or "lime hating" are ill-named; the determining factor is really either an intolerance of high levels of calcium, or a special need for iron, manganese, boron, or other trace elements made unavailable by the alkaline conditions. It is claimed in America that such acid loving plants as azaleas, gardenias and rhododendrons grow readily on alkaline soils if supplied with chelates that enable them to take up the necessary iron and presumably manganese from the soil.

There are two ways of dealing with soil reaction. One is to change it to the value desired; the other is to accept it and grow plants that tolerate it. Actually a compromise is usually the most comfortable arrangement. Reduction of soil acidity is quite easy: all that has to be done is to add lime or ground limestone, and the local agricultural or horticultural adviser can usually say how much is needed. Where a large quantity is required on a light soil it should be given in several small doses as the loss by leaching increases with the size of the dressing. Overliming a sandy soil may do damage in several ways; it is less harmful on a heavy soil.

It is much more difficult to increase soil acidity: i.e. to lower the pH. Direct addition of nitric or sulphuric acid is completely effective but very dangerous. Addition of aluminium or ferrous sulphates is safer: these decompose in the soil with formation of sulphuric acid; long continued use of sulphate of ammonia (but not of superphosphate) has the same effect. Addition of sulphur is the most convenient method; it is oxidised by some of the soil bacteria—a really remarkable group that can obtain their energy this way and seem to tolerate, or manage to avoid, the effects of the acid they produce.

There is no fixed optimum pH value for particular plants or organisms but there is a certain range outside which they do not grow well if at all. For common agricultural crops the safe ranges are:—

pH Values

Above 7	Lucerne, sugar beet, barley, wheat, potatoes.
7 *to* 6	Wheat, potatoes, red clover, turnips.
6 *to* 5	Potatoes, white clover, oats, rye, timothy, lupins.
Below 5	*Agrostis canina, Festuca ovina, Anthoxanthum odoratum.*

On heaths and hill pastures Calluna, Erica, the wiry-looking Nardus, *Deschampsia flexuosa*, Vaccinium, are all tolerant of pH values 4 and under.

EFFECT ON PLANT GROWTH	pH scale	EFFECT ON PLANT FOOD AVAILABILITY
Many flowering shrubs do well, but not the heather group, azaleas, rhododendrons, lupins or most lilies.	7·5	Reduced availability of phosphate, potassium, manganese, iron, and other trace elements except molybdenum. Plants liable to chlorosis
	7·0	
Optimum for most plants.	6·5	Maximum availability of mineral nutrients, including trace elements.
	6·0	Phosphate fixed by soil; other mineral nutrients potassium, calcium, magnesium and trace elements suffer loss by leaching. Bacteria affected more than fungi. Nitrification reduced.
	5·5	
Many plants suffer from acidity, roots short, stubby, often fanged. Lupins do well but most clovers do not.	5·0	Phosphate largely unavailable.
	4·5	
Heaths and other moorland plants, azalea, rhododendrons do well but many flowering plants fail.		Most mineral nutrients become much more soluble and liable to be washed out. Soluble aluminium appears in harmful quantities. Bacteria and many soil animals seriously affected.

Soil reaction. pH 7 represents neutrality, lower values acidity and higher values alkalinity. Acidity becomes 10 times more intense for each fall of one unit: thus it is 10 times more intense at 6, 100 times more intense at 5, and 1,000 times more intense at 4, than at 7

FIG. 11

Changes in soil and plant relationships as soil reaction changes from alkalinity to acidity of increasing intensity

The effects of soil reaction are well seen in grass fields where competition with neighbouring plants curtails the extent of adaptability that would otherwise be shown. Clovers in particular are crowded out from acid grass fields.

A large variety of garden plants flourish at pH values 6·5 to 7, i.e. neutrality and slight acidity. Of garden plants rhubarb is one of the most tolerant of acidity though it will grow equally well on a calcareous soil; strawberries and tomatoes also tolerate a wide range of reaction (pH 4·8—7·0). Cabbage, cauliflowers, beet and asparagus are more sensitive (pH 6·5–7·5); celery, lettuce, onions rather more so (pH 6·5–7·0). Among flowers, roses, primulas, schizanthus and hydrangeas are the most tolerant; the pH may range from 4·8 to 7·9; in the case of hydrangeas the flower colour is red when the pH is high and blue when it is low.[1] Ericas and similar plants, rhododendrons, azaleas, gardenias do best at low pH, probably 4·8 to 5·8. Chrysanthemums and lilies do best on acid soils (pH 6·0–6·5); *Lilium regale*, however, tolerates higher values. For sweet peas, geraniums and pelargoniums the soil should be neutral or slightly alkaline (pH 7 or over).[2] In cool wet conditions plants are more tolerant of acidity than in dry, and also on soils well treated with farmyard manure.

Soil organisms are sensitive to the soil reaction and where the limits of tolerance of parasitic or disease-causing organisms do not fully coincide with those of the host plant it is possible to effect control by changing the soil pH. Potatoes will grow on soils of pH values 7·6 to 4·5 or less, their optimum pH ranges from about 4·8 to 5·6; but the actinomycete that causes scab grows only over the range 7·6 to somewhat below 7; by keeping well below this particular range the potatoes remain free from attack. The plasmodium that causes Finger-and-toe in turnips is active when the pH is 6 or less but not at values above this; but turnips grow well up to pH 6·6, which gives them an ample margin of safety.

Looking back over this whole problem of soil control the outstanding feature has been the remarkable extent to which the chemist and the engineer are beginning to dominate the position. The engineer has produced implements that cultivate the soil and manipulate the plants

[1] If available aluminium is present. E. M. Cheney showed that this interesting change is the result of aluminium uptake, which is possible only in acid condition. (Journ. Royal Hort. Soc. *62*, 304 (1937).

[2] Prof. H. W. Miles of Wye College has kindly given me these figures.

in ways that would otherwise require higher degrees of skill and crafts-manship than are now generally available (happily there are still exceptions). The chemist has discovered and learned to make substances that effect the most remarkable changes both in the plant and in the soil. Some of these will change vegetative shoots to flowering shoots; others will hasten or retard flowering and fruiting so that harvests can be staggered, others again will produce fruit without the need for pollina-tion. Soil once broken up can be changed to a good crumb structure by chemical conditioners, its reaction can be adjusted to any desired point, its deficiencies in plant food can be corrected chemically, and chemical agents are available for the destruction of undesirable soil organisms and pests, weeds and other unwanted plants.

These chemical agents are becoming increasingly selective and therefore more numerous. The engineer is steadily working towards more automatic devices. Automation may yet invade the farm, com-mercial market garden and nursery even though an escapist reaction may keep it out of the private garden. An electronic controller pro-grammed to respond to changing weather and soil conditions as revealed by self recording instruments, and to changing plant conditions recorded by photographic devices operating light cells, may yet by remote control send the proper cultivating implement to the proper spot and direct its operations, select the proper chemical agent for each particular purpose and direct a discharge of the proper amount to the proper place; never itself making a mistake but being eternally watchful to correct those of any human intruder who thinks he knows better. Let us hope that there will always be some place left

> "*Where the sound of living water never ceaseth*
> *In God's quiet garden by the sea,*
> *And earth, the great life giver, increaseth*
> *Joy among the meadows like a tree*".

CHAPTER 8

THE SOIL AND THE LANDSCAPE

IN THIS CHAPTER I shall try to show how the various properties of soil have influenced the natural features and the utilisation of the land, and played a dominating part in determining the landscape of the countryside. A simple and easily recognisable grouping of the soils has been adopted: for the more elaborate groupings of the Soil Survey their Reports must be consulted.

THE SANDY SOILS

The sandy soils of England owe their distinctive features to the fact that their mineral particles are in the main relatively large and almost entirely composed of inert silica, which contains no plant food, has no colloidal or base exchange properties, and consequently little power of retaining water or fertilizers. They are therefore very liable to dry out in times of low rainfall, or to be leached out where rainfall is high: they are picturesquely described as being always hungry or thirsty. They always contain, however, small proportions of other substances: organic matter, oxides of iron and aluminium, clay minerals and others; some of these greatly affect the character of the soil.

The rainfall largely determines the properties of a sandy soil. It washes out nitrates so easily that plants may fail to obtain sufficient for full growth; farmyard manure has special value because it renews the supplies. High rainfall leaches out so much of the calcium and magnesium that the soil becomes acid. At a further stage the acidity becomes more pronounced and much of the iron, aluminium and manganese are also leached out from the surface soil but, by a process not fully understood, they are precipitated lower down. Three layers are often formed. The uppermost is black or brown and strongly acid: it consists chiefly of plant material in process of decomposition; below this is the leached layer of sand which, having lost its iron and manganese, is white; then comes the layer in which these bases and also alumina have been precipitated, giving the brown colour; lower still is the original rock material. (not yet changed). These soils are called by their Russian name Podzols which means "ash"; they are common on sandy heaths or in coniferous forest. The white layer which may be 4 or 5 inches thick is desperately poor and very unrewarding to the cultivator. The layer in which the iron and other bases are deposited may set to a hard pan which hampers root development and the downward passage of water. Air is thus excluded, causing asphyxiation of the plant roots and driving some of the soil organisms to get their oxygen from nitrates, ferric and manganic oxides or from sulphates, with resulting loss of nitrogen and formation of ferrous and manganous compounds and of sulphides and sulphuretted hydrogen all of which are toxic to plants.

Even when uncultivated, sandy soils are rarely left untouched: their herbage is grazed either by sheep or till recently by rabbits. Where grazing has been light there is much heather (*Calluna*) and heath (*Erica*); where it has been heavier this gives place to low growing grasses. There is a considerable variety of short-lived plants, often for a week in spring only when the soil is moist, including Ladies' bedstraw, harebell, tormentil and others; *Holcus mollis* and sweet vernal grass are common, and ragwort often abundant, attractive with its yellow flowers till the cinnabar moth strips its leaves and there remain only gaunt naked stems. Conifers flourish and where there is moisture, willows. On acid sands there are sessile oaks, wavy hair grass (*Deschampsia flexuosa*) and sheep's sorrel. Where, as on the Bagshot sands, partings of clay hold up water the untidy oak and scrub vegetation give at first sight the impression of a clay landscape; closer inspection shows marked differences however. At the other extreme, in dry poor conditions there are many lichens, species of *Cladonia*.[1] Even small differences in elevation may lead to considerable variation in moisture content of the soil with corresponding variations in the flora.

The poverty and ungrateful nature of the sandy soils caused great stretches to be left uncultivated to the present day: Breckland, in Suffolk and Norfolk, the Surrey and Dorset heaths are examples.

[1] For full accounts of the floras of the soils dealt with in this chapter see A. G. Tansley, *Britain's Green Mantle* (1949), London, Allen and Unwin, and *The British Islands and their Vegetation* (1939) Cambridge Univ. Press.

Their inhabitants, the "heathen" were poor benighted people existing miserably and precariously on patches of soil better than the surrounding land but unattractive to good farmers. Cobbett in his *Rural Rides* of 1822 is vitriolic on the subject of these "rascally heaths"; particularly Hindhead, "certainly the most villainous spot God ever made".

Reclamation of the sandy soils began in earnest in Norfolk in the 18th Century, first at Raynham near Fakenham by Charles, 2nd Viscount Townshend in 1730 after his retirement from politics, and 46 years later by Thomas William Coke, afterwards Earl of Leicester, at Holkham. The method was to add marl, which fortunately occurs locally, so increasing the capacity for holding water and plant food and for forming crumbs that would not easily be blown away. Organic matter was added frequently as animal manure and root residues: in order to provide this numerous farm animals were kept and sequences of crops grown to ensure ample supplies of food for them, particularly for folded sheep (p. 171) and for cattle in the yards: leguminous crops, turnips, swedes, and later mangolds were very important. Coke widened the range of fodder crops so reducing the risk of the disastrous hungry gap that would occur if one of them failed. This still remains the best way of dealing with a sandy soil. Marling is not considered feasible nowadays but addition of organic matter is, especially since the wider use of lucerne.

One of the most striking light land reclamations of modern times has been Lord Iveagh's great enterprise of bringing into cultivation at Elveden in Suffolk 9,000 or 10,000 acres mostly waste, some with a small output of rye and sheep. (Plate XIV, p. 199). The land was surrounded by a rabbit proof fence (a costly operation but myxomatosis had not yet appeared) then ploughed so as to bury the wild vegetation, a process easier for bracken than for heather as it grows on deeper soil; then as the soil is both acid and poor, chalk (locally available in quantity) and appropriate fertilizers were given. Two cereal crops were grown to provide an immediate cash return. In the second of these a mixture of lucerne and cocksfoot was sown: both are deep rooting and drought resistant; the mixture was left down for three to five years during which time it much enriched the soil to some depth in useful organic matter. Next came a cereal crop, then sugar beet, finally another cereal crop in which lucerne and cocksfoot were sown, thus starting a second cycle. The lucerne and fodder grains and straws are fed to dairy cattle, beef cattle and sheep; large quantities of food are obtained on land that had previously produced little but game birds and rabbits.

An alternative method of utilisation is to take advantage of three properties of sandy soils of great importance to nursery men and market gardeners. Their lack of stickiness makes them easy to cultivate even when wet and has earned for them the name "light", though bulk for bulk they are the heaviest of all soils. Their permeability to air enables plants to make great masses of fibrous roots, and if of deep rooting habit to go down several feet: sugar beet roots at Woburn have been traced to a depth of nearly five feet. (Plate XVIa, p. 207). Good root crops can consequently be grown: potatoes are important in the north west of England, and sugar beet in the eastern regions. Productiveness is greatly enhanced if the water table is so near the surface that roots can get into the capillary fringe.

Their low power of holding water allows the sandy soils to warm up rapidly in spring and early summer, so that plants start early and grow quickly, producing early flowers and fruit. Ripening also is early: indeed it is often too rapid to give high quality in certain crops, for example malting barley.

Some of the best known nurseries and market gardens in the world are on the light sands of the south east of England and the western Netherlands; no other soils can produce shrubs and young trees with such beautiful roots, or such early fruits and vegetables. Wells may have to be sunk and installations set up for watering the plants in periods of drought; large expenditure is necessary on fertilizers and manure, constant action against weeds is necessary. The risks are great, but so may be the rewards. This method is practicable only where ample water and supplies of organic manure are available, and also a market serving a sophisticated *clientèle* able and willing to pay the necessary high price for early produce.

Where no such facilities are available the land can be planted with trees: conifers are among the easiest to grow as they tolerate both poverty and acidity, and once past the nursery stage they grow satisfactorily; large areas have been planted by the Forestry Commission. Their most destructive pests are heedless people who start fires: many thousands of acres of heath and forest perished in the dry spring of 1956; fire brigades were called out to deal with some 140,000 acres, the highest number since collection of statistics began in 1946 and there were many fires for which they were not called.

A garden on a light sand may be managed by either of the above mentioned methods: the soil may be treated with marl or clay and large quantities of organic manure so as to make it physically more like a loam, or it can be treated ecologically, and set with plants that grow well in such a habitat. Beautiful specimens of ornamental conifers large or small, rhododendrons, azaleas, brooms, heaths and heathers of many kinds can be grown; and among flowers lupins, lilies, tulips and other bulbs, many alpines, and with attention to manuring and liming many herbaceous plants and shrubs. Liming if necessary should be in small doses: it is not necessary or desirable to aim at neutrality: a pH of 6 to 6·5 is usually satisfactory. Potash seems to be particularly necessary for plants on sandy soils: salt used to be applied in the old days. Fertilizers should be given for each crop: one cannot count much on residual effects.

Grass may be more difficult to establish, especially in the dry eastern counties: cocksfoot may survive but some desirable species are liable to suffer from drought and be crowded out by weeds. Lawns and greens may present special difficulties which however can be referred to the Bingley Research Station or to one of the highly expert seed firms. Owing to their low content of organic matter and of clay the upper and the lower layers of sandy soils may differ but little. A gale on May 5th 1955 blew away the surface soil and the crop of a Breckland farm and it was feared that the land might long suffer. But it did not: sugar beet sown immediately afterwards grew quite well. (Plate XV, p. 206). Sand dunes can be fixed by marram grass; thereafter a more varied flora can become established. The most spectacular utilisation of sand dunes is in the western Netherlands; the upper part is carted away for road making or building and the base is used for growing intensive market garden and nursery crops by what is almost soil-less culture. (p. 166).

THE LOAMS

There is little to be said about the loams except to praise them. They are very productive; the landscape looks clean, neat and tidy, well set off with majestic elms; farms and villages have a prosperous look. Rain leaches some of the calcium carbonate and other basic material from the soil, but much of it is taken up by the broad leaved trees and other plants and returned to the surface again; this constant

Plate XIII. *a.* Repeated mowing of grass in well-kept orchard, Cambridgeshire.

b. The Tartarian Lamb: a herbivorous plant according to John Hunter (1729)

Plate XIVa. Sandy heath near Dunwich, Suffolk.

b. The heath reclaimed: Lord Iveagh's enterprise at Elveden, Suffolk.

By Courtesy of Lord Iveagh

circulation prevents the soils from becoming very acid. Drainage is good throughout their depth: it is faulty only if some underlying material prevents the removal of the water. In consequence air penetrates easily, the iron is fully oxidised giving the soil a fairly uniform red or reddish brown colour whatever the colour of the parent material. The good aeration causes roots to develop well and also the various members of the soil population: earthworms and other animals actively mingle plant residues with the soil. As a result of these various conditions there are no very clearly marked horizons in the soil profile. There is practically no waste land on the loams: their natural fertility is enhanced by fertilizers, lime and proper crop rotations. They should be jealously conserved as our best source of home grown food. (Plates XXIII and XXIV, pp. 226 and 227).

CLAY SOILS

The properties of clay soils most affecting plant growth are their stickiness when wet, the minuteness of the pore spaces, the tendency of the soil crumbs to disintegrate, and the swelling and contraction that follow wetting and drying, often tearing roots in summer and forcing bulbs and stones out of the ground after frosts in winter. Their stickiness makes ploughing so difficult that in the old days three or four horses might be needed to draw a single furrow plough: land was characterised as "three horse" or "four horse", and the soils were described as "heavy" in contradistinction to the "light" sandy soils.

As a result of the small pore spaces water is held with such force that much of it cannot drain away and even plants cannot extract it. Air cannot readily enter, root growth is much more restricted than on sandy soils and there is no massive production of fibrous roots. Owing to the high content of water the soil warms but slowly in spring, sowing is late, so also is the harvest, often to the detriment of the quality of crops such as malting barley. If all has gone well yields may be good, but there are many possibilities of failure. The nature of the subsoil plays a very important part. Some clays, like the Clay-with-Flints of the Home Counties, have cracks and fissures in the subsoil down which the water can soak away to a water table or a porous stratum of sand or chalk (Plate XVI*b*, p. 207) others have a more compact subsoil which holds up the water and necessitates artificial drainage. Frequently the lower layers contain much more of the finest material than the surface soil, presumably the result of washing down by the rain.

In consequence of these difficulties the clays have had a very chequered career. They were originally covered with forest, chiefly oak,[1] ash and thorn with hazel and elder undergrowth; in places, hornbeam, as in the London Clay cappings of the high ground north of London; their clearance must have been a severe task for our Celtic and Saxon forefathers. Their cumbersome implements and slow moving oxen could not always finish cultivation during the most favourable period and a wet season was often disastrous. On the other hand "Drought never bred dearth in England" as the old proverb went; hard clods could be broken with a mallet (Plate XII, p. 179) and the three common crops, rye on poor acid soils, wheat in better conditions, and beans, all formed strong deep roots and in a dry season could search the soil for water; moreover they were not too seriously damaged by the cracks that formed in a dry season. The land was laid in ridges and furrows to ensure drainage for part at least; in a wet year the furrows carried many weeds and in a dry year they might be baked very hard; in neither case would they bear much crop.

The first great improvement came when drainage began to change from a craft into a science, but there were many catastrophies during the process.[2] Cultivation, however, remained very slow and costly: only during a run of good seasons or of high prices could the clay soils be cultivated to advantage. Much land had to be left untilled; the roads were so desperately bad that they had to be wide, there was neither time nor energy for keeping the hedges neatly trimmed and they grew up bedraggled and unkempt. Quite commonly the arable land lay fallow one year in three, partly because the weeds from the preceding crop provided a little food for the animals, partly to allow more time for the accumulation of nitrates and for making a good seed bed for the important wheat crop.

The best periods for the clay soils were during the Napoleonic wars,

---

[1] *Quercus robur*, pedunculate or common oak, which carries its leaves direct on the twigs and its acorns on stalks; not to be confused with *Q. sessiliflora*, sessile oak, which has its leaves on stalks and its acorns direct on the twigs, and which is more common on the acid siliceous soils of the northern and western counties and Wales.

[2] *Talpa, or The Chronicles of a Clay farm,* by C. Wren Hoskyns, 1852, gives an entertaining account of the struggle of an enterprising farmer wanting to use improved drainage technique with his old craftsman who had "been a-draining this forty year and ought to know summut about it".

the 1850's to 1870's, the 1914-18 war and the 1939-45 war: when the high prices ended the clays ceased to be arable and were either sown down to grass, or more often, allowed to tumble down to grass. Usually this resulted in a mass of almost useless weeds; some, such as ox-eyed daisies and quaking grass found in the drier places, are attractive enough to us but not to the cattle that have to feed there; others like yellow rattle are parasitic on the roots of grasses. The outlets of the drains were no longer kept clear, the land became waterlogged, covered with surface creeping vegetation, black bent (*Agrostis vulgaris*), buttercup and others, with sedges and rushes in wet places; elsewhere brambles, thorn and others made a kind of scrub bushland.

Two types of management are now practised. The soil may be sown with a mixture of perennial grasses and clovers which are left down indefinitely. Its texture gradually improves (p. 31) and with proper additions of lime and fertilizer the herbage grows satisfactorily. Alternatively the land may be kept in arable cultivation: this has been greatly facilitated by several notable developments. The crawler tractor, multiple tillage implements and improved harvesting machinery, have so expedited operations that they can be carried out directly favourable conditions set in, and may be completed before they change: speed of working is essential in the management of clay soils. Drainage schemes can now be planned on sound engineering lines and carried out with tiles or moles as circumstances require.

Varieties of cereals have been bred specially suited to clay soils and the seed can be treated with a chemical agent to destroy wire worms, one of the most serious soil pests of ploughed-up clay pastures. Sugar beet, introduced into England in the 1920's, has proved a valuable crop, and modern implements with their greater power and higher rate of working have enabled farmers to grow Brussels sprouts, potatoes and other crops formerly confined to the loams. The fertilizer requirements are usually relatively simple. Phosphates are necessary to stimulate root development and nitrogen to increase leaf growth; potash is usually present in sufficient quantity in the soil for cereals and grass but is needed for potatoes, sugar beet and many garden plants. Lime, chalk or ground lime stone are required on acid soils: the pH may be raised to 7 or even higher; over-liming is less harmful on a clay than on a sandy soil.

But the risks of bad seasons always remain, and, as clay soils are the last to come into arable cultivation when the need arises, so they are among the first to go out and be laid down to grass when the need has passed. This laying down is, however, very different now from what it was. Special varieties of grasses and clovers have been selected and bred at the Plant Breeding Station, Aberystwyth (Plate XVII, p. 210) and by some of the great firms of seedsmen, and tested at the Grassland Research Institute at Hurley, Maidenhead; suitable mixtures can be sown giving good yields of nutritous herbage for farm animals. Dairy or beef cattle are the most suitable animals, the choice depending on the preference of the farmer and the location of the farm; sheep are less suitable because a clay pasture harbours liver fluke and predisposes the animals to foot rot.

The change from grass to arable farming and vice versa is easier in principle than in practice. Arable farming necessitates heavy expenditure on implements, fertilizers, seeds, often also on buildings for storage, and cottages to house the additional workers; the implements are very costly and cannot be stored for long periods as they deteriorate and quickly get out of date. Grassland farming if done properly involves expenditure on fencing and laying on water into the different fields, though admittedly much cheaper and far less productive methods are possible and even usual.

In 1914 farmers could not unaided change from grass to arable and substantial subsidies had to be given; in the severe depression of the 1920's a large part of the capital thus injected into the farms was lost and land tumbled out of arable cultivation, some farmers who patriotically tried to continue went bankrupt, and much land became derelict. When the war of 1939 came it was necessary once more to provide subsidies, this time much higher, in order to get the land back into tillage. The problem of organising clay farms in such a way that they can switch over from grass to arable and back again without this enormous loss of capital has yet to be solved. It still remains true for the farmers on the clay lands that the change from prosperity to adversity can come with disastrous suddenness. As a result of these fluctuations in fortune the clay regions have never been as well groomed as most of the English countryside had been before the postwar conditions set in; consequently the clay landscapes are among our untidiest with their rough roadside verges, straggling bedraggled hedges, broken down ditches, trees and copses scattered apparently indiscriminately. (Plate XX, p. 217).

But unlike man's greater untidyness this has its attractions: stately oaks, often of great antiquity; the village churches many of which go back to the 13th Century or earlier; the village names, commonly perpetuating those of their Saxon founder, all tell of a stream of life that has flowed steadily if inconspicuously throughout the ages, retaining much of its ancient individuality as a result of the badness of the roads and the lack of transport until the motor car came in some 50 years ago. Much of the best in the English character was nurtured in the villages on the clay lands.

The slipperiness and softening of clay when wet may cause serious disturbance of overlying beds. Small slips are common in clay valleys, but periodically dramatic land-slides occur. Gilbert White describes one at Hawkley near Selborne in 1774,[1] another of 1839 can be seen at Lyme Regis. A slip caused in a different way occurred on Folkestone Warren in 1915 disrupting the railway for 1½ miles: it resulted from the gradual softening and breakdown of the almost rock-like Gault clay when it becomes thoroughly wetted.[2] (Plate XVIII, p. 211).

Clay soils containing a large amount of silt have all the disadvantages of a heavy clay soil without the advantage of a stable subsoil with drainage fissures such as the true clays often possess; natural drainage is therefore impeded. They have also much less capacity for base exchange. In general they are more difficult to drain artificially than clays and the silt has an unfortunate way of blocking the drains. They form a glazed cap after rain, will not usually carry a tractor and they are difficult to work.[3] They can, however, be lightened by treatment with large dressings of farmyard manure—50 tons or so to the acre. Only amateur gardeners can afford such quantities, crops are too late for commercial growers. For farmers these soils are best laid down to permanent grass, but deterioration of the herbage sets in sooner or later and after a time breaking up and reseeding may be necessary. If the soil structure has been improved in the meantime a few arable crops can be grown before laying down again.

A number of modern housing estates are on clay soils, the more desirable sands being already occupied. Those round London are not

---

[1] Letter 45.

[2] Studied by A. H. Toms: *Folkestone Warren Landslips*, Railway paper No. 19, Inst. Civil Engineers 1946: continued by A. M. M. Wood 1940–50 and by N. E. V. Viner-Brady 1948–54: Railway Papers Nos. 56 & 57 respectively. (Inst. C.E. 1955).

[3] Not all silty soils are so difficult. The soil of University Farm, Aberystwyth, contains 44 per cent of silt and 22 per cent of clay, but presents no special difficulties of tilth formation or drainage. It contains, however, 20 per cent of coarse sand.

usually difficult for garden making: the London Clay is perhaps the worst. The Clay-with-flints is responsive and has good natural drainage: it is likely to need lime unless it has been well chalked in the past. The Chalky Boulder Clay also is responsive and not usually difficult. Further afield the Lower Lias Clay is in places truculent, the Oxford Clay is more manageable. Usually a preponderance of elm is a good sign: fine oaks may indicate a stiffer clay. Many shrubs including roses do well once established, also carnations, sunflowers and other sturdy herbaceous plants. Much depends on the persistence of wetness during the winter, and a process of trial and error is necessary to discover the most suitable plants. The gardener is at a disadvantage compared with the farmer that he cannot usually do much in the way of drainage. The new soil conditioners facilitate the making of a seed bed.

THE CHALK AND LIMESTONE SOILS

These are characterised by their relatively high content of calcium carbonate, the most dominating of all the soil constituents because of its profound effects in determining the properties of the soil and the flora it can carry. The chalk region forms a great stretch of country in the south and east of England: the main block starts in Dorset and Wiltshire; one section stretches across Hampshire where it forks: a southern limb goes through Sussex to Beachy Head forming the South Downs, and a northern one crosses Surrey and Kent to the East Kent coast forming the North Downs. The other section starting from Wiltshire strikes north east across Berkshire, Buckinghamshire (the Chilterns) and Hertfordshire then northwards, ending up as the Wolds of Yorkshire. The chalk is soft and somewhat soluble in water impregnated with carbon dioxide; as a result the outlines of the landscape are smooth and rounded: the escarpments and the sides of the coombs are, however, often steep.

Rain readily soaks in as the chalk is very porous, springs break out at the foot of the Downs fertilizing the lower land (Plate XIX, p. 216) and leading to the growth of a line of ancient and prosperous villages. The region of the Upper and Middle Chalk is an important source of water for the towns. Some of the streams run only when the local water table has risen sufficiently high, they become dry when it falls; these are the "bournes". This marked porosity is an important factor in determining the vegetation characteristics of the soil. An even more important factor is the calcium carbonate itself which operates in three ways: it supplies large quantities of calcium to the plant, it neutralises organic and other acids, and it throws out of action various metallic compounds some of which are essential in small quantities as iron, manganese, zinc and copper but all of which are harmful in larger amounts.

Plants vary in the extent to which they can tolerate these various effects: some are adversely affected and can no longer hold their own in the competition for a place in the soil, others are unaffected or even benefited by the conditions and prosper. The resulting flora thus comprises two groups of plants: those for which an ample supply of calcium carbonate is essential for full growth: the calcicoles, and those for which it is not essential but which tolerate it sufficiently to be able to stand up against competition from other plants.

Beech is one of the most characteristic trees of the chalk in southern and eastern England: it forms dense woods on the slopes of the valleys and the escarpments and its canopy so closely shades the ground that little undergrowth is possible. In many places the soil is no more than an inch thick, but the trees send their roots deep into the chalk. These woods provide raw material for some important rural industries, the best known is chair-making in the Chiltern region. Yew characteristically accompanies the beech, and also holly, wild cherry or gean, and in places white beam. Further west ash takes the place of beech; they are everywhere competitors but here the higher rainfall and larger amount of impurity in the chalk apparently shift the balance of competition in its favour. In more open conditions scrub develops; in the south east it includes a number of flowering shrubs, dogwood, spindle tree, wayfaring tree, privet, buckthorn, hazel, elder, field maple and the climber, traveller's joy. Till recently rabbits had severely kept these down, but now they are springing up again. No other kind of soil shows anything like so rich a variety of shrubs.

The woodland plants include wood anemone, cuckoo pint, wild strawberry, yellow deadnettle, wood spurge, sanicle and mosses on shallow soils and steep slopes, and dog's mercury on deeper soils and gentler slopes: this, however, is very aggressive and excludes many other plants. All these can complete their growth in the beechwoods before the canopy closes in. Where there is neither wood nor scrub the Downs are covered with a beautifully springy turf, close cropped for centuries by sheep and rabbits. It contains an amazing variety of plants. Some are there because they can tolerate the dryness, such as sheep's fescue, wild thyme and milkwort; others because they must have alkaline conditions: salad burnet, rock rose, and lesser scabious. Some of the plants are now becoming rare: *Anemone pulsatilla*, but some attractive orchids remain; the bee, the butterfly, even in one place, the lizard. The commonest grasses are the sheep's and red fescues, brome, the oat grasses and brachypodium; the latter is very aggressive. A. G. Tansley described the grassland of the chalk downs of south-eastern England as "one of the best characterised and also one of the most attractive of our semi-natural grassland communities".[1]

It is not clear why chalk soils should carry this marvellously varied flora. The simplest explanation appears to be that many different plants can grow there but the conditions do not allow any to grow vigorously enough to crowd out its neighbours; any seed that happens to fall can produce a seedling that will have a chance of life if it can tolerate the conditions.

A curious mixture of calcicoles and calcifuges occurs in places where the calcium carbonate has been washed out from a thin upper layer of the soil and the decomposing plant residues become acid. This allows the growth of shallow rooting acid-tolerant plants such as sheep's fescue and even in places *Calluna* and *Potentilla erecta*, typical heath plants. Below this shallow layer the soil is calcareous, and deeper rooting calcicoles establish themselves; rock rose, burnet, orchids and others.

The undulating land on the chalk plateau has a deeper soil especially in the bottoms where the soil has been carried down from the higher ground. Some of the soil, however, is not derived from the chalk at all but from later formations; this is often acid and therefore carries a different flora including oak trees and foxgloves, beech trees also will grow, but as at Burnham Beeches, they are not happy: they are crooked and not to be compared with the majestic specimens on the chalk.

The chalk lands have been farmed for some thousands of years. The first cultivators were the Neolithic peoples who came about 2500 B.C. bringing with them wheat (and some of its weeds) and cattle, they were followed about 1800 B.C. by Bronze Age people who set up Stonehenge; later came Celtic peoples who brought primitive ploughs;

[1] *Britain's Green Mantle* already quoted.

Plate XVa. A 40 m.p.h. gale removes the surface soil (May 5th 1955) carrying away the crop. *b*. Nevertheless a good crop of sugar beet was obtained later (*p*. 198)

Plate XVIa (left). Roots of sugar beet traced for nearly 5 ft. in sandy soil. Woburn Experimental Farm (*p.* 197). (*H. H. Mann*). *b (right),* Profile of Rothamsted clay soil to a depth of 4½ ft. showing fissures down which water can pass (*p.* 199). (*V. Stansfield*)

other Celtic invaders about 480 B.C. brought in iron; their small, roughly square fields can still be traced by aerial photography; still later about 75 B.C. came the Belgic peoples with bigger implements requiring long rectangular fields, also still traceable; shortly thereafter history begins.

After the Black Death of 1389 had almost depopulated great stretches of the countryside, the chalk and limestone regions of the south became great sheep ranches, wool being the product as in Australia and South Africa today. Much of the wealth thus gained was spent on building the beautiful churches and later delightful manor houses, farms, and choice little towns like Burford and Chipping Campden that make the Cotswold region one of the most attractive in the country. It still possesses gems of craftsmanship which once they fall to decay could no longer be replaced, the urge and the taste having gone.

Sheep long continued to be the mainstay of the chalk and oolite countryside. In the 17th Century John Evelyn in his Diary describes Salisbury Plain as "that goodly plaine, or rather Sea of Carpet, which I think for evenesse, extent, verdure, innumerable flocks, to be one of the most delightful prospects in nature". Two generations later Daniel Defoe in his tour of the chalk regions[1] was almost lyrical about them, "the most charming Plains that can any where be seen, the vast Flocks of Sheep[2] . . . a great Part of these Downs comes by a new Method of Husbandry, to be not only made Arable, which they never were in former Days, but to bear excellent Wheat and great Crops too, tho' otherwise poor barren Land, and never known to our Ancestors to be capable of any such Thing".

The "new Method" was the folding of sheep on the arable land (p. 171); by the early years of the 20th Century it had developed into ingenious alternations of grain crops with fodder crops skilfully adjusted to avoid failure of food supplies for the sheep whatever the season. But as already stated, in recent years rising costs of labour and the disinclination of the young men to become shepherds have made the system uneconomical, and dairy cattle have gradually displaced the sheep. The output of food per acre has fallen, but the system is

[1] *A tour of Great Britain*, 1724–1727.
[2] He was told that within six miles of Dorchester (Dorset) there were 600,000 sheep. Apparently so much sheepbreeding induced peaceful habits for he goes on: " Here I saw the Church of *England* Clergymen, and the Dissenting Minister or Preacher drinking Tea together, and Conversing with Civility".

sufficiently profitable to survive. The love of sheep, however, never died in the villages, and some of the Oolite farmers are keeping more of them but making more use of grass, so saving labour. Modern insecticides and medicines have greatly reduced disease troubles, so lessening the labour and anxiety of shepherding, and it is not uncommon for the farmer himself to undertake the work. Chalk and oolite soils are peculiarly well suited to sheep because of their dryness and the short habit of growth of the vegetation.

New varieties of wheat and barley have proved eminently suitable for these soils and give much higher yields than the old sorts. Atle, a Swedish wheat, was bred for other conditions and rejected: fortunately it was tried on the chalk land and proved a resounding success.[1] Dr. Bell's Proctor barley also does well. Weeds in considerable variety flourish on chalk soils: charlock in particular was a great pest in barley, but can now be completely eradicated by hormone herbicides.

Gardeners on chalk soils are in the happy position of being able to grow a great variety of shrubs and flowering plants, but they may have trouble with some of the plants they particularly want to grow, especially fruit trees which are liable to lime-induced chlorosis, a disease that is difficult to cure, though chelates promise to solve this problem. The soil is apt after heavy rain to dry in hard steely lumps, not easy to break down to a fine tilth except after frost or at a certain stage of moistness. Organic manures are very effective. The soil population is so varied and so numerous that the organic matter is quickly decomposed: large quantities are used by intensive cultivators. Potash is especially important as a fertilizer, and where the rainfall is fairly high, phosphate also. A soil is not necessarily a chalk soil because it lies on the chalk, however: it may be one of the acid patches and must be treated accordingly.

Peat, Moor and Fen Land Soils

These are characterised by a high amount of organic matter which completely masks the effect of the mineral constituents; their properties depend on the way in which this organic matter accumulated. It

[1] Till the recent outbreak of myxomatosis rabbits did so much damage to winter wheat on chalk soils—sometimes taking as much as one quarter of the crop—that farmers had to grow the lower yielding spring wheats instead. They are now reverting to winter wheat, and fortunately the French produced some very good sorts such as Cappelle Desprez, which have become very popular.

persisted because the dead vegetation remained on the surface of the soil instead of being drawn in by earth worms and other animals; in consequence it did not undergo the usual drastic decomposition. This happened because the soil was waterlogged. Two types of decomposition took place. Where the water was well charged with calcium carbonate the acids produced by the decomposition were largely neutralised, but where the water came mainly from the rain or from mountain streams in granite or sandstone regions it contained little or no basic material and in consequence the acids persisted and of course determined the type of micro- and macro-populations, the changes they could effect, and the plants that could grow. The acid material is called acid peat and the more nearly neutral material, lowland or fen peat.

The acid peats are formed mainly on high ground in the high rainfall regions of the north and west of Great Britain. In presence of so much water the conditions are very acid, anaerobic and unsuited to most bacteria; there is little or no nitrification, and only those plants can grow that can utilise ammonia, amide or amino-compounds, or like heather, have associated mycorrhizal fungi. (p. 76). These conditions greatly restrict the flora: and as usual in such cases small local variations of position often have marked effects. Under the wettest conditions, with high rainfall, perpetually moist atmosphere, and little drainage the dominant plants are Sphagnum moss and beaked sedge (*Rhynchospora alba*); in somewhat drier but still very wet conditions cotton grass and deer sedge (*Scirpus caespitosus*); in still drier places the leafy grass *Molinia,* and in the driest situations, e.g. on a hummock, heather and bilberry. Associated with the wetter conditions are bog asphodel, cross leaved heath (*Erica tetralix*), and cranberry.[1] Trees are generally ruled out by high winds and by the general unsuitability of bog conditions; they grew, however, in the earlier stages of development of the bog, and trunks and roots can often be found, preserved by the lack of micro-organic activity. Such decomposition as takes place is effected by fungi, but it is only partial, and stops at the formation of a black compact mass of humic material, slimy when wet, hard when dry, and possessing many of the properties already described on p. 23. Its high power of holding water results in much swelling, and it has happened in Ireland and occasionally in England that a bog at the top of a hill has overflowed into the valley after

[1] A full and interesting account of bogs, moors and fens is given by A. G. Tansley in *Britain's Green Mantle* already quoted.

unusually heavy rainfall, doing great damage. The water from a peat bog contains soluble humic acids; it is brown in colour and very soft; it dissolves lead from pipes and is therefore unsuitable as water supply to a town until the acid is removed.

The black material is used as fuel; it is cut in spring, stacked for drying during the summer and carried away in autumn. The more fibrous material lying above this black stratum is used in stables as litter because of its high power of absorbing liquids.

Several kinds of bog are recognised. In very wet conditions the whole surface of the land may be covered by a "blanket bog" as in parts of Connemara, Western Mayo, Western Scotland and Dartmoor. Under drier conditions the bog is more localised and tends to grow more in its interior than at its outside fringe; it is called a "raised bog". Few of these now remain in England, they having been drained and reclaimed, but a good one persists at Tregaron in Cardiganshire because a sill of rock where the river flows out from the bog holds the water up.

Reclamation of the peat bogs has been frequently attempted but is made extremely difficult by the persistent high rainfall. Some success has been claimed for efforts to start rain after a dry spell but no one has ever succeeded in inducing it to stop. The first step is to provide drainage by open trenches, though the removal of the water may be a major engineering undertaking, involving the deepening of the river bed or the removal of some bar such as a sill of rock which is keeping the river water above the level necessary for proper discharge. As the land becomes drier heather begins to dominate the vegetation: it can be burned periodically and the young shoots can be grazed by sheep. As grazing continues grasses come in and gradually take possession; cattle can then be introduced, but productivity is low. The process is slow, but inexpensive once drainage is done. It can be hastened by ploughing and sowing a suitable seeds mixture when the land has become sufficiently firm to bear the weight of the implements, but this work is not free from risk; it has happened that the plough has dropped into one of the holes that are so common on bogland and so difficult to see; once in there is great difficulty in getting it out. English experience on Dartmoor has been assembled at Dartington Hall, Totnes, and at the Seale Hayne Agricultural College, Newton Abbot; and Irish experience at the Department of Agriculture, Dublin.

Where adequate drainage is practicable a system can be adopted

The Field

Plate XVII. Welsh Plant breeding station, Aberystwyth. *a.* Artificial pollination to produce new strains of clover.

b. Field trials of new strains of oats.

D. J. Griffiths

Plate XVIII. Landslide in Folkestone Warren caused by breakdown of the Gault clay when wet: disruption of the railway line, 1915.

which has been used successfully in the Netherlands. The brown
sphagnum layer at the surface is set aside; the black peat is all taken
for fuel; the base of the bog, usually sand in the Netherlands, is levelled;
the sphagnum layer is spread evenly on it and more sand from canals
and ditches is spread on that; good dressings of fertilizer are given.
The high rainfall and acid conditions rule out many crops, but certain
grasses, potatoes, oats, and rye will grow and a suitable rotation can
be drawn up. Grass and cattle rearing make the best combination.
The moorlands in Britain, however, lie high, and the farm houses are
necessarily isolated; the work is heavy and unrewarding, and most
people find the life depressing. It is not feasible to adopt the "suitcase
farming" practised in parts of Canada, as in the Peace River Settle-
ment: the very pleasant summer is spent on the farm but the very trying
winter in more agreeable places. Grain, however, is the product there,
and it leaves no responsibilities once harvest is safely over.

Where water can escape from the hill lands another type of flora
develops: grasses of various sorts, gorse, bracken and other plants.
Farmers judge soil by the vegetation it carries: Welsh farmers who have
done much reclamation of hill lands have a saying "under bracken
gold, under gorse silver, under heather copper". The Scottish version
is somewhat similar except that it goes "under heather famine". There
is much truth in the old saying. Bracken flourishes only on deep rather
dry soils, gorse tolerates poorer conditions, and heather requires high
acidity in order that its mycorrhizal fungus *Phoma* can develop; it also
indicates poverty and will grow on very poor sandy soils.

The fen soils are completely different from those of the moorlands.
They are under a much lower rainfall and are situated at lower levels:
the great fens in the region of the Wash are almost at sea level. As in
the case of the moors drainage is the chief problem, but its purpose is
to prevent flooding when high tides bank up the water in the rivers
which meander through them. Originally they were a great waste of
scrub vegetation inhabited by wild fowl, fish and eels. Some draining
was done in medieval times, but the great scheme was started in 1626
by a celebrated Dutchman, Cornelius Vermuyden, with the financial
backing of the then Duke of Bedford and his fellow "Adventurers"; the
work has continued ever since. The rivers were dyked, main drains
and a network of subsidiaries were cut, pumps were installed to lift the
water into the rivers, and their operation so regulated that the water
level neither falls too low nor rises too high.

The removal of the water allowed air to enter and this, with the mineral salts and particularly the calcium carbonate carried in by the ground water enabled an active population of micro-organisms to establish itself, with the result that decomposition of the organic matter proceeded rapidly. This, together with the shrinkage consequent on drying and the removal of free water, caused a sinking in level of the land. When Whittlesey Mere was drained in 1850 the owner, W. Wells of Holme, had the happy idea of driving three posts through the peat into the Gault clay below and making the top of each flush with the surface of the soil: later an iron post replaced one of these. By 1870 the post was nearly 8 feet out of the ground,[1] and since then shrinkage has continued: by 1956 the level had fallen 12 ft. 2½ ins. (Plate XXI, p. 220). This rapid rate of decomposition results in considerable production of nitrate, and as the skilful regulation of water level always ensures an adequate water supply, the fen soils have high productive power.

The region is free from the wetness and sunlessness of the moors and the vast expanse of sky and the wide stretch of level land give it considerable charm: there are no difficulties in finding farmer settlers.[2] The soils lack phosphate: many farmers indeed use no other fertilizer than superphosphate though the Rothamsted workers found that potatoes responded to fuller manuring. Cereals do not, however: indeed the crops are often so heavy that they become badly lodged; this happens so frequently that farmers have learned well how to deal with them. Modern stiff strawed varieties stand up better.

An old method of improving the soil was to dress it with the underlying Gault clay brought up by trenching; this however, is now too costly for general use. Very large crops of wheat, oats, potatoes and sugar beet are grown, often considerably exceeding those obtained on the ordinary agricultural land outside the fen. The quality of the

---

[1] The measurements up to 1932 are recorded by Gordon Fowler in *Geogr. J.* (1933), 81, 149. The 1956 value was kindly taken for me by Dr. E. C. Childs. The values are:

	1848	1860	1870	1875	1892	1932	1956
		ft. ins.	*ft. ins.*	*ft. ins.*	*ft. ins.*	*ft. ins.*	*ft. ins.*
	—	4 9	7 8	8 2	10 2	10 8	12 2½
Average rate of shrinkage inches per year		4.8	3.5	1.2	1.4	0.15	0.8

[2] An interesting account of the rehabilitation of a derelict fen farm is given by Alan Bloom in *The Farm in the fen* (1943) London. Faber & Faber.

potatoes is not particularly good, there being a tendency to blacken after boiling, also the percentage of sugar in the sugar beet is below the average on the agricultural land. These defects, however, are more than compensated by the high yields. Two special crops are very successful: celery and mustard: the latter has contributed greatly to the wealth of Norwich and made the fortune of the house of Colman.

So productive are the fen soils that practically all are now in cultivation; bus services have been organised to take the children to school and the women into the surrounding towns for shopping. Wicken Fen is owned by the National Trust and has been kept as a Nature Reserve; it is under the control of Cambridge University and as the public are excluded it retains its wild flora and fauna undestroyed. Both differ greatly from those of bogs and moors. Reeds (*Phragmites*) are common in the wetter parts along with reed mace (*Typha angustifolia*), bulrush, and *Molinia;* in the drier parts woody plants develop: the low growing bog myrtle and creeping willow, guelder rose, buckthorn, alder and birch. A wooded area is locally called "carr": left completely undisturbed it would develop into forest.

Little patches of bog vegetation occasionally develop. Sphagnum moss may settle on clumps of *Molinia* or *Cladium* above the water level; they receive only rain water and no mineral salts, consequently on decomposing they form acid peat. On this may grow sundews, cotton grass and other plants typical of peat bogs, but not seen on the true fen.

High-Lying Soils

Altitude has no direct effect on soil but it acts indirectly through its influence on rainfall and drainage. The effect of high rainfall in leaching out the basic material, with successive production first of acid soil, then finally of podzol conditions, has been discussed on pages 194 and 195. In addition on sloping ground there is a steady downward movement of soil particles which pile up at the bottom of the slope and gradually get worked down to a level by further action of rain, of soil animals and of cultivation. These soil particles also intercept much of the mineral material washed out from the higher levels thereby becoming richer in bases and consequently less acid; this and the greater depth makes the soil a better medium for plant growth than the land higher up, and these flats or "bottoms" as they are commonly called in the south are often sown with grass in wet districts or with cereals and fodder crops in the drier chalk regions (Plates XIX and XXVI). So it happens that in travelling down a long and fairly steep slope the soil at the top may be distinctly acid, carrying a calcifugous flora, it becomes less and less acid in going downwards, and at the bottom it may be neutral or nearly so, with corresponding changes in the flora.

High lying level land receiving no seepage or other water from neighbouring hills may be kept in cultivation provided the rainfall is not too high. On the East Yorkshire wolds cultivation is carried on at altitudes of 800 feet, but the rainfall averages only about 27 to 30 inches yearly. Further west, altitudes and rainfall are much higher: the Derbyshire and Yorkshire moors can, however be cultivated up to about 1,000 ft. under a rainfall of 45 to 50 inches as in the Hope and Edale Valleys, and to still higher levels in the rain shadow west of Sheffield where the fall is 35 to 38 inches only.[1] Above these levels of altitudes and rainfall the soil becomes far too acid for cultivation even when sufficiently level: it becomes covered with peat which in places is eroded by gullying leaving very uneven surfaces. Reclamation of lower lying hill land has been attempted in Montgomery and elsewhere, but the need for constant replenishment of lime and plant nutrients makes cultivation expensive, and the isolation and climatic conditions over much of the year make life unattractive so that little permanent success has been attained. Many signs of abandoned cultivation can be seen in the hill districts of the western counties.[2]

Nevertheless the hill farms of Great Britain make a substantial contribution to the total agricultural output of the country. (Plate XXII, p. 221). The results of an enquiry by B. K. Davidson and G. P. Wibberley of Wye College[3] are given in Table 19 (opposite).

[1] Prof. A. R. Clapham informs me that much wheat is grown above 1,000 ft. in this region; potatoes, swedes, and kale up to 1,260 ft.; oats to 1,310 ft., while grass land can be ploughed up and reseeded (with perennial rye grass, cocksfoot, and clovers) up to 1,400 ft.

[2] Factors influencing the upper limit of cultivation in different parts of England and Wales are discussed by L. Dudley Stamp in *The Land of Britain, its Use & Misuse,* (1950). London. (Longmans).

[3] Studies in Rural land use. No. 3 (1956) Wye College, Kent.

Plate XIX. The rolling chalk Downs, Sussex. Bottom land cultivated.

Plate XX. The untidy clay landscape, near Ettington, Warwickshire.

TABLE 19
Agricultural production from hill farms of Great Britain, 1952/3.

MILLION ACRES

	MOUNTAIN AND HEATHLAND	ALL GREAT BRITAIN
Total area	16·5	56
Used for farming	13·5	45
Crops and grass	2·2	28·8

VALUE £M

Gross agricultural output ..	47	1,202
Net ,, ,, ..	40	1,019
Including Government subsidies	7·6	312

CHAPTER 9

SOIL ANALYSIS AND SOIL SURVEY

A CENTURY AGO chemists thought the whole problem of plant nutrition was settled. Sir Roderick Murchison, President of the British Association in 1856, proclaimed in full confidence that if agriculturists "wish to solve their doubts respecting the qualities of soils, or the effects of various manures upon them, our chemists are at hand." Looking back at the methods used it is extremely difficult to see how they could possibly have given much useful advice to good farmers. The revival of agricultural science in the 1890's and early 1900's produced greatly improved methods but also showed their inadequacy, bringing out clearly that the relation between the chemical composition of the soil and the growth of the plant is affected by its environmental conditions, particularly the water supply, soil structure, and reaction. The final arbiter is the plant itself, and the effect of any treatment can only be assessed with certainty by actual trial. It is not necessary, however, to make trials in every garden and every field: results obtained at a carefully chosen centre can reasonably be expected at other places with similar soil and rainfall conditions. Farmers have long recognised this. The Agricultural Surveys of the different Counties of the United Kingdom made by the first Board of Agriculture in the early years of the 19th Century contained soil maps based on simple classifications such as sands, loams, clays, peats and fens.[1]

The increasing need for agricultural improvement called for better methods of soil classification and for systematic soil surveys. The first on an extensive scale was by A. D. Hall and E. J. Russell and covered the counties of Kent, Surrey and Sussex. This happens to be a well defined geological region free from complications due to drift and the soils are in general derived from the underlying rocks. The geological formations therefore formed a satisfactory basis for the broad classifi-

[1] The 8vo. series. An earlier 4to. edition starting in 1793 had no soil maps.

cation and subgroups were based on the soil texture as shown by mechanical analysis.

The geological basis by itself, however, proved unsuitable in many districts north of the Thames. Over large areas the soil is formed from drift and not from the underlying rock, and although in many cases the drift may have come from similar rocks it has frequently been derived from some other formation a considerable distance away. Further, the same rock can give rise to one kind of soil under one set of climatic conditions and to a totally different soil under another set. Climatic factors must therefore be taken into account. This was first emphasised by the Russian workers, and their investigations of the relationships involved constitute perhaps their best known contribution to agricultural science. Meanwhile in the United States the classification was based mainly on texture as revealed by mechanical analysis.

Present day systems of classification combine all three factors and are based on climate, parent rock and texture. This has involved a complete change of outlook. The older surveys were frankly intended for use by farmers and gardeners and were therefore confined to the top 9 or 18 inches of soil, the depth below which cultivation operations do not usually extend; the modern survey deals with soils as distinct natural objects, and the ultimate basis of the classification is the manner in which they have been formed. The whole profile of the soil has therefore to be studied, from the surface down to the immediate parent material, and this may be a depth of 3 or 4 feet. It is usually divided into three horizons, the upper, called A, characterised by the presence of humus from dead plant residues; a lower horizon B, which contains the clay, silt and other material washed out from A; and a third still lower, the immediate parent material called C; usually these are further subdivided.

Soils with profiles of similar type are put into the same major group: of these there are about half a dozen in the British Islands. Two, the Brown earths (p. 35) and the podzols, (p. 195) are widespread: the former in well drained regions originally broad leaved woodland, the latter on heaths and in natural coniferous forest. Both are naturally acid, the former slightly, the latter usually strongly. Calcareous soils are found on the chalk and other limestone formations: in the former case they are whitish or grey, in the latter red or brown; they are neutral or slightly alkaline. Peat and fen soils (p. 208) are more local. Cutting across these divisions is another due to lack of aeration where drainage is impeded: any of the above groups may thus become a gley (p. 17), either from surface water or from ground water.

These major groups are divided into series in each of which the profiles are similar: the soils are derived from the same parent material and have been subjected to similar weathering influences. Each series is generally named after the place where it was first seen. Further subdivision is then made on the basis of differences in texture, whether sand, loam, clay or something intermediate: about a dozen divisions are recognised by British soil surveyors.

The tests applied to the different horizons of the profile are the colour, texture, structure, and simple physical properties such as consistence and porosity which can be assessed on the spot by observation or some simple test as rubbing between finger and thumb: for type profiles, however, these properties are measured in the laboratory. No two profiles are absolutely alike, but different surveyors agree reasonably well in classifying a given set of soils. Where the conditions over a given area have varied continuously, as for instance where two parallel mountain slopes enclose a valley, no sharp natural division into series is possible, but the system can be treated as a whole: it is then called a Catena.

Once a satisfactory method of classification was worked out, it became possible to organise a systematic soil survey for the whole country. This was started for England and Wales in 1939 under the direction of G. W. Robinson but it was held up during the war: it was restarted on a broader basis in 1946 under A. Muir at Rothamsted. The Survey for Scotland is centred at the Macaulay Soil Institute at Craigiebuckler; it is carried out on lines similar to those adopted in England, though a different grouping of the soil series has been adopted. In both countries the reconnaissance survey is made on a scale of 2½ inches to the mile and the detailed survey on a scale of 6 inches to the mile. In both countries the field survey is made on a scale of 2½ inches to the mile and published maps are on a scale of 1 inch to the mile.[1]

The advantages of a reliable soil map are many. Soils shown to be similar may be expected to respond similarly to fertilizer and other treatments and to be similarly utilisable. Experimental centres can

[1] Further details of Survey procedure are given in *Soil Survey of Great Britain*, Rpt. No. 1. of the Soil Survey. (Research Board of the Agric. Research Council (1950). H.M.S.O).

be set up and the results applied with a higher degree of certainty than is otherwise possible. Areas can be delimited which should at all costs be saved for food production. Claims have been made that housing estates produce food of the same value as that of the farms or market gardens displaced. There is a large element of special pleading about this as the comparison is between the wholsesale price of farm products and the retail price of those from the garden. In any case gardens and allotments produce neither bread, butter, milk, cheese, beef, mutton or pigmeat, and only few surplus eggs. The national diet without these things would be meagre and dreary in the extreme.

CHAPTER 10

WHAT WE OBTAIN FROM OUR SOIL

THE MOST IMPORTANT product of our soil is food for ourselves and our animals. The value of the food produced can be assessed in several ways; the commonest are the content of calories and the market value. Calories can be measured and expressed numerically, but they represent only part of the value of the food; proteins, vitamins, mineral substances and palatability are equally essential factors, and as it usually happens that foods rich in these are more expensive, market price gives a measure of desirability if not of physiological value.

The last war had a profound influence on food production in the United Kingdom. Before 1939 the average number of calories in the human food produced per acre of cultivated land (crops and grass) was about 550,000; in the record year of agricultural output (1943/4) it had been raised to 850,000, and by 1950 it was about 770,000.[1] (Table 17, p. 225). The average calorie consumption per head is a little over 1·1 millions per annum: 100 acres therefore supply the needs of 67 persons, i.e. 1½ acres per head. The 31 million acres of cultivated land in the United Kingdom thus provide for 21 million people, 41 per cent of our population of 51 millions: the remaining calories have to be imported. The pre-war production sufficed for only 31 per cent of a smaller population: there has therefore been a marked improvement in output.

This production of 41 per cent of our calories does not mean that we produce 41 per cent of each of our foods: of some we produce much

---

[1] James Wyllie (1954) *Land requirements in the production of human food,* Wye College. The figures include some calories derived from imported feeding stuffs but they disregard the output of gardens, allotments and holdings under 1 acre the produce of which however, with the exception of potatoes is more important for the supply of vitamins and other accessory substances than of calories. It is impossible to evaluate these quantities precisely: they work in opposite directions and tend to cancel out.

more, of others less. Recent figures for the percentages produced at home are:—

Milk and potatoes	100	Wheat and flour	28
Carcass meat	67	Sugar	21
Bacon and ham	46	Fats and Oils	16
		Butter	10

Owing to the relatively high price of the various meats the home produced food comes out higher in value than in calories:—

Percentage of our food produced in the United Kingdom[1]

	Prewar	1950-2	1953-4
Calorie basis	31	41	41
Value at constant prices	36	50·3	50·5

During this period the total production went up more than the figures indicate, the population having increased. The net output has been:—

1936-9	1951-2	1952-3	1953-4
100	148	152	156

As every housewife knows, the average Englishman is obstinately conservative in the matter of diet, and the detailed figures for consumption per head of the 14 chief items of food in 1955 are almost identical with those for the average of the pre-war years of 1934-8, excepting that for milk and potatoes there had been an increase of 25 per cent, and for fish an increase of 18 per cent. Consumption of butter was somewhat down, but that of margarine was up and the two together showed little change; flour was slightly down. This stability of demand conduces to a stable pattern of food production on our farms, varied only by the effects of competition from overseas producers.

Of Britain's 31 million acres of crops and grass about 40 per cent are in cereals, roots and other tillage crops; and 60 per cent are in grass, rather more than two-thirds of which is permanent and the rest is in

[1] Parl. Quest. 5th Dec. 1955.

Plate XXII. Hill Sheep Farming. Auchallater Farm, Glen Clunie, Aberdeenshire
(*P. K. McLaren*)

leys of varying duration,[1] followed by a sequence of arable crops. The 40 per cent of our land in tillage crops supplies 60 per cent of our home production of calories, while the 60 per cent in grass supplies only 40 per cent; the grass, however, enriches the soil and improves its structure so benefiting the tillage crops. (p. 29). The low calorie production from the grass land is due to the low efficiency of the animals as producers of human food. Dairy cows head the list, transferring to their milk 20 per cent of the calories and 23 per cent of the protein in the food they eat: the milk of course was intended for the calf but we have no scruples about taking it, either slaughtering the calf or rearing it apart from its mother. Pigs are almost as efficient in transferring the calories of their food into flesh and fat suitable for human beings, but they are only half as efficient as protein converters: for this purpose good laying hens are the best. But beef, our favourite meat, is very extravagantly produced, containing only 7 per cent of the calories and 10 per cent of the protein the animals had consumed: mutton is very little better, the conversion factors being 8 and 10 per cent respectively. This low ability to produce human food has been discovered empirically by many of the peoples of the world. Poor overcrowded and peasant nations have a largely vegetarian diet and no cultivated grassland. Pigs and poultry may be kept where animal foods are not forbidden by religion as for instance in China, but they have to feed on scraps. Only the richer nations consume much beef or mutton. We ourselves during the war had to curtail our production of animal foods: by growing human food crops on land thus liberated we were able to raise our total calorie production from 16·7 million million before the war to 26·3 million million in 1943/44. (Table 18, p. 226). The calories supplied by livestock fell by 18 per cent, but those supplied by crops rose by 57 per cent.

The increased outputs recorded above have been achieved in spite of a loss of more than half a million acres[2] taken permanently out of agricultural use; they represent the marked increase in yields of crops and livestock resulting from the improved methods and materials now available. Before the war the average yield of wheat was 17 or 18 cwt.

[1] One to five years or more, the shorter periods in the dry eastern counties where the leys consist of rye grass, cocksfoot and white clover, and the longer periods in the wetter north and west where timothy and meadow fescue do better.

[2] The Ministry of Agriculture figures are somewhat smaller, but R. H. Best in *The Major Land Uses of Britain* (Wye College) shows that their estimates are too low.

Plate XXIa. Shrinkage of fen after draining in 1850. The surface was then level with the top of the post (*p.* 212). (*The Nature Conservancy*)

Soilless culture of carnations . 166). Horticultural Experimental ation, Aalsmeer, Netherlands.
(*L. A. J. Mascini*)

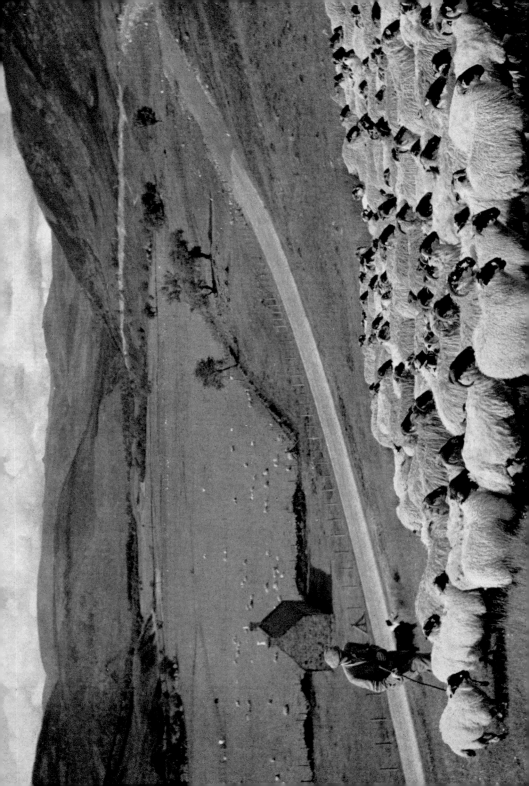

per acre: for the years 1953-8 it was 24·7 cwt.; other yields have also risen.

This marked increase in production from our soils is very gratifying, but the cost has been colossal. Fertilizer consumption has trebled, tractors have increased five-fold, heavy expenditure has been incurred on drainage, liming, grubbing unwanted hedges and waste scrub, providing water for the animals, improvement of buildings and other necessities. Farmers themselves could not possibly have found the money, and heavy taxation prevents the landlords from doing so. Large State subsidies are necessary each year for ploughing up grass land, draining, purchase of lime and fertilizers, and to make up the difference between the cost of production and the selling price of the food. For the year 1958-9 farm subsidies were £73·1 m. and Price Guarantee Payments £191·1 m.; the total Government expenditure on Agriculture was £358·7 m.[1] against £314·4 m. in 1956-7. The subsidies must be on a level sufficient to keep the weaker farmers in business and they are therefore on a generous scale for the better farmers (who incidentally are often on the better land): the money, however, is well used and is spent on increasing or modernising their equipment. These subsidies are likely to be a permanent feature of our agriculture, and may indeed have to be increased as the change in our system of land tenure and the social structure of the countryside continues.

Considerable further increase in output is physically possible. The present average yield of wheat is 24·5 cwt. per acre but a good farmer expects 30 or 35 cwt.; the record, however, is 70 cwt. per acre. For potatoes the present average yield is 8 tons per acre, but a good Lincolnshire farmer expects 13 or 14 tons and the record is 27 tons per acre. When the factors responsible for these high yields are known and when it becomes possible to reproduce them on our farms, great increases in output may be expected.

The question is often asked: can we ever become self-supporting? The answer turns on what precisely is meant. Our population is increasing and our cultivated land is steadily shrinking in area: our present cultivated area of 0·8 acre per head may soon fall to 0·7 and while further reclamation of heath and moorland is physically possible it is costly and most unlikely to be self supporting. If we

[1]This total included: Ministry of Agriculture £17.6 m; various services to agriculture £10.9 m., including £2.3 m. for education and certain research and £4.1 m. for the Agriculture Research Council, the main research organizing body.

adopted a largely vegetarian diet, using as food only such animals as could be maintained on farm wastes and byproducts such as straw, tail corn, milling offals, sugar beet residues, and the minimum of grass needed for soil fertility purposes, we should need less land than at present. However it was achieved complete self sufficiency would involve drastic changes. We know how much it has cost to raise the pre-war 31 per cent of self sufficiency to the present 41 per cent: what it would cost to raise this to 100 per cent must be left to the imagination.

This increasing cost of higher output is largely inherent in crop production. G. V. Jacks has pointed out[1] that agricultural development in all countries passes through three phases. The first is one of low level self sufficiency by primitive methods necessitating shifting cultivation: this is still found in parts of Africa. Later comes the more settled stage, also of self sufficiency, when the soil becomes gradually exhausted and the surface layers may be lost by erosion. This stage is wide-spread at present and accounts for much of the world's hunger. Finally there comes the stage of soil rehabilitation when improvements are effected which, however, necessitate the importation from outside of ameliorating agents that have to be paid for. This is possible only where industry has sufficiently developed to find the necessary capital. Farmers themselves cannot do this because agriculture is an individualist enterprise, it makes little appeal to joint stock companies because its operations are not under complete control and its risks are incalculable. A successful farmer may absorb holdings of some of his less competent neighbours but usually the added units are again separated when he retires.

Modern aids to higher output are apt to introduce new problems not easily solved. Modern chemicals control pests and diseases, but they may also upset Nature's equilibrium of wild life with consequences yet unknown; moreover they may get into the plants and so into our food: insecticides have already been found in milk. The increased use of modern fertilizers has raised important deficiency problems and the closer packing of plants and animals on the land has encouraged the development and spread of diseases. New and speedier machines have increased the risks of injury to the workers: some 25,000 accidents a year are reported on British farms. Yet progress cannot be halted: these modern appliances are absolutely indispensable. The husbandmen's path will always be beset with difficulties: Virgil's words remain eternally true:

[1] *Advancement of Science* (1956). 13, 137–47.

. . . . The Father of Agriculture
Gave us a hard calling; he first decreed it an Art
To work the fields, sent worries to sharpen our mortal wits
And would not allow his realm to grow listless from lethargy.

Like other advanced countries the United Kingdom is covered with
a network of agricultural research, education and advisory services
staffed with men and women seeking to make the soil give up its
secrets. Inventors and manufacturers devise improved appliances,
while farmers and farm workers use the new knowledge and devices to
enhance their output. Short of a world upheaval there seems no cause
for anxiety for the future.

TABLE 17

Acreages and production in human food: United Kingdom.

	TOTAL AREA	AGRICULTURAL USE		NON-AGRICULTURAL
		CULTIVATED CROPS AND GRASS	ROUGH GRAZINGS	
Million acres ..	60·15	31	17	12

	CROPPED LAND, MILLION ACRES				TOTAL PRODUCTION, MILLION TONS			
	1936/9	1943/4	1949/50	1954/5	1936/9	1943/4	1949/50	1954/5
Cereals ..	5·30	9·56	8·02	7·75	4·44	8·64	8·03	8·06
Potatoes ..	·72	1·39	1·31	·94	4·87	9·82	9·03	7·32
Sugar beet	·33	·42	·42	·44	2·74	3·76	3·96	4·52
Fruit & veg.	·58	·72	·86	0·8	2·82	3·8	3·19	(a)
Other crops	2·0	2·36	2·1	1·9				
Total tillage	8·9	14·5	12·7	11·83				

(a) not available.

TABLE 18

PRODUCTION OF CALORIES FOR HUMAN CONSUMPTION (J. WYLLIE):

	MILLION MILLIONS			PER CENT OF TOTAL		
	1936/9	1943/4	1949/50	1936/9	1943/4	1949/50
Cereals ..	2·94	11·17	6·66	17·6	42·5	27·9
Potatoes ..	2·50	4·78	4·51	15·0	18·2	18·9
Sugar beet ..	1·53	2·10	2·22	9·2	8·0	9·3
Fruit & veg.	·48	·66	·62	2·9	2·5	2·6
All crops ..	7·46	18·71	14·00	44·7	71·2	58·7
Milk ..	4·83	4·88	6·21	28·9	18·6	26·1
Eggs ..	·43	·22	·48	2·6	0·8	2·0
All meat ..	3·97	2·46	3·14	23·8	9·4	13·2
Total livestock	9·23	7·57	9·84	55·3	28·8	41·3
Total calories home produced	16·69	26·28	23·84	100	100	100

Plate XXIII. Fertile Loam, Wye. Land to be treasured.

W. Holmes

Aero Pictorial Lt

Plate XXIV. Intensive farming: fruit, hops, crops, on fertile sand, Mockbeggar, Beltringe, Kent.

BIBLIOGRAPHY

GENERAL

HALL, A. D. (1931). The Soil; an introduction to the scientific study of the growth of crops. 4th Edition revised by G. W. Robinson, London, John Murray.
> The first edition appeared in 1903 and was the first English book to present the subject in its modern form.

JACKS, G. V. (1954). Soil. Edinburgh, Nelson. pp. ix & 221.

BEAR, F. E. (1953). Soils and Fertilizers. 4th ed. New York, Wiley: London, Chapman & Hall. pp. xiii & 420.

ROBINSON, G. W. (1949). Soils, Their Origin, Constitution and Classification. London. T. Murby. pp. 573.

LEEPER, G. W. (1948). Introduction to Soil Science. Victoria, Melbourne University Press. pp. 222.

RUSSELL, E. J. (1940). A Students book of soils and manures. (3rd Ed.) Cambridge Univ. Press. pp. viii & 291.
> Soil conditions and Plant growth. 9th Ed. revised and rewritten by E. W. Russell. (1961). Longmans, Green & Co.
> An advanced treatise dealing at length with the scientific foundations of the subject.

GODWIN, H. (1956). History of the British Flora. Cambridge, University Press.

MANLEY, GORDON (1952). Climate and the British Scene. London, Collins New Naturalist Series.

SOIL COMPOSITION AND PROPERTIES

BEAR, F. E. (Editor) (1955). Chemistry of the Soil. New York, Reinhold Publishing Corp., London, Chapman & Hall. pp. 373.

BAVER, L. D. (1956). Soil Physics. (3rd Edn.). New York, John Wiley & Sons. pp. 489.

GRIM, R. E. (1953). Clay Mineralogy. New York, McGraw-Hill Book Co. pp. 384.

BRAGG, W. L. The atomic structure of minerals. (Cornell Univ. Press).

wos—Q

SOIL BIOLOGY
WAKSMAN, S. A. (1952). Soil Microbiology. New York. John Wiley & Sons, London, Chapman & Hall.
SANDON, H. (1927). Composition and distribution of the Protozoan Fauna of the Soil. Edinburgh. Oliver & Boyd.
KEVAN, D. K. McE. (1955). Soil Zoology, London, Butterworth. (1962). Soil animals, London, H. E. & S. Wetherby.
KUHNEELT, WILHELM (1961). Soil biology with special reference to the animal kingdom. Trans. by Norman Wallace, London, Faber & Faber.
GARRETT, S. D. (1956). Biology of root-infecting fungi. Cambridge, University Press.
GOODEY, T. (1963—rev. ed.). Soil & Fresh Water Nematodes, London, Methuen.
DARWIN, C. (1881). The formation of vegetable mould through the action of worms, with observation of their habits.
HOSKINS, C. P. (1945). Of Ants and men. London. Allen & Unwin.
ELLIS, A. E. (1926). British Snails. Oxford. Clarendon Press.
MATTHEWS, L. HARRISON (1952). British Mammals. London, Collins New Naturalist Series. pp. xii and 410.
SMART, J., and TAYLOR, G. (1953). Bibliography of the Key Works for the identification of the British Fauna & Flora. (2nd Edn.). London. Systematics Association. Gives lists of the most important works on the various members of the soil population.

FERTILIZERS
MILLAR, C. E. (1955). Soil Fertility. New York. John Wiley & Sons, pp. 436.
SMITH, A. M. (1952). Manures and Fertilizers. Edinburgh, Nelson. pp. 275.
WALLACE, T. (1951). The diagnosis of mineral deficiencies in plants. (2nd Edn.). London. H.M.S.O.
IGNATIEFF, V. and PAGE, H. J. (1958). F.A.O. Agricultural Studies No. 43 F.A.O., Rome (obtainable from H. M. Stationery Office, London).

WATER & SOIL
ADDISON, H. (1955). Land, Water and Food, London. Chapman & Hall.
NICHOLSON, H. H. (1953). The principles of Land Drainage. (2nd Edn.). pp. 163.
KENDALL, R. G. (1950). Land Drainage. London. Faber & Faber. pp. 133.

SOIL & LANDSCAPE

STAMP, L. DUDLEY. (1946). Britain's Structure & Scenery. London, Collins New Naturalist Series. pp. xvi & 255.

(1955). Man and the Land, London, Collins New Naturalist Series. pp. xvi & 272.

HOSKINS, W. G. (1955). The making of the English Landscape. London, Hodder & Stoughton. pp. 240.

(1949). Midland England. London. Batsford. pp. viii & 120.

TANSLEY, A. G. (1949). Britain's Green Mantle, London, Allen & Unwin. pp. xii & 294.

SOIL SURVEYS.

HALL, A. D. and RUSSELL, E. J. (1911). A Report on the Agriculture and Soils of Kent, Surrey and Sussex. London. H.M.S.O. pp. viii & 206.

KUBIENA, W. L. (1953). The Soils of Europe. London. T. Murby. pp. 318.

UNITED STATES DEPARTMENT OF AGRICULTURE SOIL SURVEY STAFF. (1951). Soil Survey Manual. U.S.D.A. Handbook 18. pp. 503.

CLARKE, G. R. (1941). The Study of the Soil in the Field. (3rd Edn.). Oxford. University Press. pp. 228.

H.M.S.O. Reports of the Soil Survey of Great Britain. London (issued periodically).

FOOD PRODUCTION.

RUSSELL, E. J. (1954). World Population and World Food Supplies. London. Allen & Unwin. pp. 514.
Describes the application of the principles of soil control to food production in various countries of the world.

(1956). The Land called me. An Autobiography. London. Allen & Unwin. pp. 286.
Includes non-technical accounts of food production problems as studied by the author in various countries.

INDEX

Vermiculite, 11
Vermuyden, C., 211
Viner-Brady, N. E. V., 203
Virtanen, A. I., 65
Volume change in soil, 42

WAIN, R. L., 187
Waksman, S. A., 71, 228
Wallace, T., 148, 156, 228
Watson, S. J., 157
Warington, Katherine, 153
Warington, R., 68-9
Warren, R. G., 26, 28
Water in soil, 39-45, 170
  control of, 177-182
  loss during droughts, 41, 43
  rate of evaporation, 41
  volume in pore space, 36-7

Weeds, harmful effects of, 176
  control of, 185-7
Weil, J., 126
Wheeler, Sir Mortimer, 5
Whistler, R. L., 32
White, Gilbert, 108, 112, 203
White grubs, 120
Wilfarth, H., 62
Wibberley, G. P., 214
Williams, E. G., 24
Winogradsky, S., 62, 69
Wireworms, 117-120
Wood, A. M. M., 203
Woodlice, 131
Wright, C. S., 179
Wyllie, J., 220, 226

ZINC AS PLANT NUTRIENT, 155